Bullen 4/21/03

MOLECULAR MECHANISMS OF IMMUNOLOGICAL SELF-RECOGNITION

MOLECULAR MECHANISMS OF IMMUNOLOGICAL SELF-RECOGNITION

Edited by

FREDERICK W. ALT
HENRY J. VOGEL

College of Physicians and Surgeons
Columbia University
New York, New York

ACADEMIC PRESS, INC.
Harcourt Brace Jovanovich, Publishers

San Diego   New York   Boston
London   Sydney   Tokyo   Toronto

This book is printed on acid-free paper. ∞

Academic Press, Inc.
1250 Sixth Avenue, San Diego, California 92101-4311

United Kingdom Edition published by
Academic Press Limited
24–28 Oval Road, London NW1 7DX

Library of Congress Cataloging-in-Publication Data

Molecular mechanisms of immunological self-recognition / edited by
    Frederick W. Alt and Henry J. Vogel.
        p.   cm.
    Includes index.
    ISBN 0-12-053750-8
    1. Immunological tolerance. 2. Cellular recognition.   3. T cells.
    4. B cells.   I. Alt, Frederick W.   II. Vogel, Henry J. (Henry
James), date.
    [DNLM: 1. Autoimmunity.   2. B-Lymphocytes–immunology.   3. Cell
Communication.   4. Immune Tolerance.   5. T-Lymphocytes–immunology.
QW 568 M7186]
QR188.4.M65   1993
616.07'9–dc20
DNLM/DLC
for Library of Congress   92-49201
                                                                              CIP

Printed in MEXICO
92   93   94   95   96   97   DO   9   8   7   6   5   4   3   2   1

Contents

3 Tolerant Autoreactive B Lymphocytes in the Follicular Mantle Zone Compartment: Substrates for Receptor Editing and Reform

CHRISTOPHER C. GOODNOW, DAVID Y. MASON,
MARGARET JONES, AND ELIZABETH ADAMS

PART III LYMPHOCYTE SIGNALING

4 T Cell Anergy

BART BEVERLY, LYNDA CHIODETTI, KURT A. BRORSON,
RONALD H. SCHWARTZ, AND DANIEL MUELLER

5 The T Cell Antigen Receptor: Biochemical Aspects of Signal Transduction

LAWRENCE E. SAMELSON, JEFFREY N. SIEGEL,
ANDREW F. PHILLIPS, PILAR GARCIA-MORALES,
YASUHIRO MINAMI, RICHARD D. KLAUSNER,
MARY C. FLETCHER, AND CARL H. JUNE

**21   Prevention, Suppression, and Treatment of
       Experimental Autoimmune Encephalomyelitis with
       a Synthetic T Cel Receptor V Region Peptide**

HALINA OFFNER, GEORGE HASHIM, YUAN K. CHOU,
DENNIS BOURDETTE, AND ARTHUR A. VANDENBARK

PART VI   PERSPECTIVES

22   Tolerance of Self: Present and Future

N. A. MITCHISON

Preface

The foundation for our understanding of immunological self-recognition, laid by the seminal discoveries of Frank M. Burnet, Peter B. Medawar, Gustav J. V. Nossal, and Joshua Lederberg, is embodied in the theory of clonal selection. The introductory chapter of this volume by Sir Gustav, written some three decades after the award of the Nobel Prize to Burnet and Medawar, summarizes the dawn of our insight into immunological tolerance, and provides an overview of research on the underlying mechanisms. Distinguishing self from nonself in the immune system, at one time a puzzling mystery in the biological sciences, is now at a stage where its molecular features are being unraveled.

The immune system presents a large number (perhaps 10^{11}) of receptors whose capacity to recognize foreign invaders and to mediate their removal is the basis of its defensive role. Clearly, the ability to distinguish proteins, nucleic acids, carbohydrates, and other organic substances, with respect to their self or nonself character, is crucial to the functioning of the immune system. Autoimmunity is normally prevented because the cells of the immune system are screened for their ability to discriminate between self and nonself through processes of clonal elimination or clonal inactivation of self-reactive lymphocytes. When the discrimination between self and nonself breaks down, antiself lymphocytes can precipitate autoimmune disease.

Parts I through IV of this volume address developments in our understanding of the molecular mechanisms of B and T cell tolerance; Part V deals with the failure of tolerance in autoimmunity; and Part VI contains the concluding chapter by N. Avrion Mitchison, which furnishes orienting perspectives and highlights new information presented in this volume. The novel findings, which Mitchison characterizes as impressive advances, pertain to the areas of B cell development and the generation of molecular diversity; V gene usage, especially from transgenes, in positive and negative thymic selection; the handling of positive and negative signals by T and B cells; anergy in postthymic T cells; the design of peptide-based therapy for autoimmune diseases; and the design of therapy with the aid of monoclonal antibodies.

It is a pleasure to acknowledge the advice and help of Dr. Charles A. Janeway, Jr., Dr. Joshua Lederberg, Dr. Polly Matzinger, Dr. Hugh O. McDevitt,

Dr. N. A. Mitchison, Dr. G. J. V. Nossal, Dr. Klaus Rajewsky, Dr. Ronald H. Schwartz, Dr. Lex H. T. Van der Ploeg, and Dr. Harald von Boehmer.

We are grateful for the continued interest of Dr. Donald F. Tapley and for the fine support of the College of Physicians and Surgeons (P&S) of Columbia University, without which this volume would not have reached fruition. This volume was developed from a P&S Biomedical Sciences Symposium held at Arden House, on the Harriman Campus of Columbia University.

This volume is dedicated to the memory of Frank Macfarlane Burnet and Peter Brian Medawar.

Frederick W. Alt
Henry J. Vogel

PART I

INTRODUCTION

1

Immunological Tolerance Revisited in the Molecular Era[1]

G. J. V. NOSSAL
The Walter and Eliza Hall Institute of Medical Research
Post Office, The Royal Melbourne Hospital
Victoria 3050, Australia

It is a very great honor indeed to contribute the introductory chapter to this volume dedicated to the memory of F. M. Burnet and P. B. Medawar, 30 years after the award to them of the 1960 Nobel Prize. As Burnet's student and successor, I also applaud the choice of J. Lederberg,[2] because his support of Burnet's theories from 1957 on was of inestimable value in giving Burnet the courage to proceed, and N. A. Mitchison, Medawar's student, to summarize and point the way to the future.

THE DAWN OF IMMUNOLOGIC TOLERANCE

For Burnet, the saga of immunologic tolerance began in 1949 (1). Stimulated by Traub's (2) finding that mice congenitally infected with lymphocytic choriomeningitis virus failed to make specific antibody and Owen's (3) observation that binovular twin cattle sharing a common placenta become chimeras, each tolerating the other's nonidentical red blood cells, he developed the

---

[1]Based on the opening address given at the P&S Biomedical Sciences Symposium, "Molecular Mechanisms of Immunological Self-Recognition."

[2]On that occasion, J. Lederberg presented an evening lecture, and N. A. Mitchison (see this volume), the concluding address.

Molecular Mechanisms of
Immunological Self-Recognition

"self-marker" theory of immunologic tolerance (1). This stated that cells and tissues bear (chemically undefined) molecules which characterize selfhood, and that if a foreign antigen were introduced into the body before the immune system had developed, the defense system would be tricked into regarding this foreign material as self, thus not forming antibody against it. What an interesting presentiment of our current knowledge of major histocompatibility complex (MHC) molecules and their role in both positive and negative selection! Burnet (4) tried to obtain evidence for acquired immunologic tolerance by injecting influenza virus into chick embryos and testing their later immune response capacity, but failed to get confirmation and temporarily let the matter drop. Medawar, nonplussed by finding that binovular cattle twins accepted each other's skin grafts, came across Burnet and Fenner's (1) monograph while browsing. He soon set about trying to induce skin graft tolerance experimentally by injecting allogeneic cells into fetal mice, and, of course, the rest is history (5).

Despite this brilliant research, the cellular mechanisms underlying tolerance lacked a coherent intellectual framework until the clonal selection theory came along. In 1955 Jerne (6) presented the natural selection theory of antibody formation, which saw macrophages as the factory for replicas of natural antibody molecules brought there because of their union with antigen. While the detailed formulation now appears clumsy, the central and deeply important point of Jerne's intervention was that the great specificity of antibodies does not have to imply instruction. There being 10^{17} antibody molecules per milliliter of serum, it would be possible to have 10^6/ml of every one of 10^{11} different specificities, the latter number being quite large enough to accommodate all that was known about antibody specificity. Talmage (7) was the first to make the connection between antibody as a natural product and Ehrlich's (8) original notion of antibodies as cellular receptors, which suggested that cells rather than serum antibody molecules might be the logical target of antigenic selection. Burnet (9) had been independently thinking along much the same lines, and reading Talmage's paper triggered his own publication of his clonal selection theory. In this first iteration, he already saw how critical it would be to have only *one* kind of receptor per immunocyte, for that would permit tolerance to be seen as the opposite of immunity. If antigen seeing "its" lymphocyte in adult life selected the cell for proliferation and differentiation to effector cell status, then antigen acting in fetal life, i.e., on immature lymphocytes, would cause deletion. Normally, of course, the only antigens present in fetal life would be self-antigens. If, for any reason, antiself lymphocytes ("forbidden clones") escaped such deletion, the stage would be set for autoimmune disease (10). As Jerne (11) wrote rather ruefully on the occasion of Burnet's 80th birthday, "I hit the nail, but Burnet hit the nail on its head."

There is one interesting sidelight to Burnet's Nobel Prize. He was, of course, a very distinguished experimental virologist and had been nominated several times by the great Sir Henry Dale. When Wendell Stanley and others were successful for basic virology, Burnet confesses he gave up hope. He certainly did not feel that the self-marker theory as such warranted the Prize, whereas he thought the clonal selection theory did and later wrote along those lines to Jerne. However, clonal selection was not generally accepted until the late 1960s, and it is doubtful whether it influenced the Nobel Committee to a significant degree. Given the entrenched nature of the direct template theory and the poor reception that the successor (12) to the self-marker theory (a book entitled "Enzyme, Antigen and Virus") received, Burnet was not at all confident about the (at the time) rather outrageous clonal selection theory. Indeed, he published it in a rather obscure journal so that if later events were to show it to be incorrect, the paper could be quietly forgotten.

ONE CELL–ONE ANTIBODY, AND IMPLICATIONS

I return therefore to the role that Lederberg (13) played. In the (northern hemisphere) fall of 1957, he spent a 3-month sabbatical in Burnet's laboratory. He had come ostensibly to work on recombination among influenza viruses, a phenomenon discovered by Burnet using incredibly old-fashioned methodology. Instead, he found the laboratory brewing with the ferment of clonal selection, first published in September 1957. He was immediately fascinated with the elegance of the idea. As a young medical graduate studying under Burnet, I suggested a way in which we could rapidly *disprove* the theory. If we could culture single cells from a multiply immunized animal and show that each cell could make two or three antibodies simultaneously (the result I predicted), clonal selection would be dead in the water. Lederberg's experience with micromanipulating single bacteria would surely allow us to micromanipulate single lymph node cells.

Lederberg dropped all other work to embrace this experiment. We debated the relative merits of two sensitive assays for antibody formation, complement-dependent lysis of erythrocytes and immobilization of motile bacteria by anti-flagellar antibody. Because of Lederberg's own experience, we chose the latter. Each of us in our own laboratories later tried to develop a single cell assay for the former as well. We failed; had we not done so, the Jerne hemolytic plaque technique might have happened 5 years earlier! However, the *Salmonella* immobilization assay served us well. There was a temporary hiccup. We needed to see whether *Salmonella* bacteria micromanipulated into a microdroplet which contained antiflagellar antibody made by a single cell would actually stop dead in their tracks. This required a microscope with a dark-field condenser, or else how would we visualize the motile bacteria? Alas, there *was* no such microscope in The Walter and Eliza Hall Institute save the sacrosanct instrument which lived in Burnet's own office! The Empire came to the rescue. Both England and Australia used very large copper pennies at that time. Josh Lederberg and I figured out that if we placed a penny in one of the filter-holders under the condenser of my old Bausch and Lomb microscope, just slightly eccentrically, we could cause a fair imitation of a dark-field effect. And so we proceeded to immunize rats with two or three different strains of *Salmonella* bacteria, and the isolated single lymph node cells in our little microdroplets always told us the same story (14): one cell, one antibody. The first step toward the eventual validation (15) of clonal selection had been taken.

Lederberg (13) soon saw a deficiency in Burnet's formulation of immunologic tolerance. Burnet concentrated on the immaturity of the individual. However, lymphocytes are born throughout the life of an animal, so what matters is the life history of each individual immunocyte. Lederberg saw each lymphocyte as moving, in its ontogeny, be it in the fetus or in an adult animal, from a stage where it was *obligatorily paralyzable* to one where it would be *inducible*. Any lymphocyte whose receptors would be directed against a "self"-antigen would, in its immature state, display those receptors first in a "sea" of self-antigen and would immediately be purged from the repertoire. If the immunocyte passed that "trapdoor," it would mature into a cell that would be "immunocompetent." The key difference between a self-antigen and a "nonself" or foreign antigen would be that the former would always be there, to "catch" the maturing immunocytes and eliminate them, whereas the latter would be pulsed in unexpectedly, perhaps tolerizing a few immature cells but really being eliminated rapidly by the majority of reactive cells that had matured beyond the trapdoor and were thus ready to form antibody. This modification of Burnet's original view accommodated medically important situations such as an immune system being reconstituted after heavy ionizing radiation. It also liberated a great deal of thinking about how the immune system could be continually renewed from stem cell sources.

It is important to recall that these pivotal insights preceded three crucial developments in cellular immunology: first, the demonstration that different immunocytes really did have different receptors (16); second, the clear delineation of the thymus and the bone marrow as primary lymphoid organs versus the spleen, lymph nodes, and other sites as secondary lymphoid organs (17, 18); and third, the division of lymphocytes into the two great families, T cells and B cells (19, 20). It is really amazing how fresh the writings of Burnet and Lederberg still appear when one realizes that all these discoveries were then well into the future.

IN VITRO LYMPHOCYTE CLONING TECHNIQUES VALIDATE REPERTOIRE PURGING AS A TOLERANCE MECHANISM; CLONAL ABORTION VERSUS CLONAL ANERGY

Technical reasons gravely delayed a direct test of Lederberg's hypothesis. Whereas antigen-binding B cells could be seen by immunofluorescence or autoradiographic techniques, it proved difficult to (a) isolate them in pure form or (b) clone them *in vitro*. We felt that only an accurate enumeration of antibody-forming cell precursors (AFCP), achieved through their cloning *in vitro* and proving that the clones made the relevant antibody, could validly permit a test of the idea that a toleragen reduced their numbers.

We succeeded in 1975 in developing a *polyclonal* system in which maturing B cells from adult bone marrow could be rendered tolerant by remarkably low concentrations of antigen (21) and in 1976 in establishing a single cell cloning system for purified, isolated hapten-specific B cells (15). Demonstration that tolerance involved a marked reduction in clonable hapten-specific AFCP soon followed (21, 22), and similar findings were reported independently by Metcalf and Klinman (23, 24). However, we were not prepared for a further finding that resulted from a careful survey of the numbers of hapten-binding B cells that could be found in the spleen after *in vivo* tolerance induction during fetal or newborn life (25). With moderate to high doses of toleragen, there was a significant though transient reduction; i.e., "clonal abortion" had been induced in the classical Lederberg sense. With lower doses of toleragen, however, hapten-binding cells could be found in normal numbers. However, the cells concerned could not be induced to form antibody in clonal microculture, whether mitogenic or antigenic stimuli were used (25, 26). It appeared that the cell had registered and stored some negative signal which ensured that it could not develop into an antibody-forming clone. So we had to devise a new term to describe this curious state: clonal anergy (25). It appeared that the "choice" between abortion and anergy depended on the strength of the negative signal. High doses of a highly multivalent (strongly cross-linking) hapten–protein conjugate led to clonal abortion; low doses, even of an oligovalent conjugate, to anergy.

This general thesis was confirmed using anti-μ immunoglobulin (Ig) heavy chain antibody as a kind of "universal" toleragen (27). It had been well known that anti-μ could frustrate the emergence of B cells and create essentially an agammaglobulinemic animal. We arranged circumstances such that small pre-B cells from newborn spleen or adult bone marrow developed into B cells in short-term tissue culture and were subsequently cloned in a second-stage culture. Moderate to high concentrations of anti-μ aborted B cell development, as expected. Much lower concentrations permitted the emergence of a normal number of B cells with a normal quantity of surface Ig receptors. However, these cells were profoundly anergic and failed to proliferate or form antibody. This further example of clonal abortion and clonal anergy as alternatives depending on ligand concentration adds to the probability that strength of signal is a major determinant between the two alternatives.

As far as the T lymphocyte is concerned, single cell cloning methods were also slow to develop, particularly before the crucial role of interleukin-2 (IL-2) was appreciated. Experiments involving bulk culture techniques or adoptive transfer were fairly evenly divided between results favoring clonal deletion as the cause of tolerance and those favoring T cell–mediated suppression. We were the first to show (28) that the injection of semiallogeneic spleen cells into newborn mice [an experimental design very similar to the original Billingham *et al.* (5) tolerance experiments] led to functional clonal deletion of cytotoxic T lymphocyte precursors. Two reasons made us believe that this occurred in the thymus. First, the kinetics favored a disappearance beginning in the thymus and noted in the spleen only a few days later. Second, specific functional deletion could be obtained *in vitro* when embryonic thymus anlagen were cocultured with fetal liver as a source of thymus-populating stem cells (29). Unfortunately, in the absence of a marker capable of recognizing specific antiallogeneic CD8$^+$ T cells, we could not distinguish between clonal abortion and clonal anergy. However, the work of Lamb *et al.* (30) demonstrated that clonal anergy could be induced in T cells that encountered antigen in the absence of costimulatory signals from accessory cells, an issue which is taken up later in this volume.

CLONAL ABORTION AND CLONAL ANERGY COME OF AGE IN THE MOLECULAR ERA

As many of the other chapters in this volume will demonstrate, the phenomena of clonal abortion and clonal anergy have been amply validated for both T and B cells by a spate of elegant experiments over the last 3–4 years. One major development has been the use of transgenic mouse technology, including the understanding of the importance of promoters which can target expression of transgenes to particular cells or tissues. Transgenic mice present two special advantages for tolerance research. The first is the capacity to have most of the immunocytes of a mouse, be they T cells or B cells, carry a receptor of a given specificity. This avoids the major difficulty imposed on tolerance research by the heterogeneity of lymphocytes. Furthermore, a toleragen (be it some soluble antigen or one on the surface of certain cells) which is the expressed product of a gene will usually be present early in ontogeny and in locations and concentrations that are held steady, as with authentic self-antigens. It is therefore a much better mimic of self-tolerance than an antigen injected artificially into a fetal or newborn animal at a single time. Another major advance has been the recognition that certain antigens react with all T cells expressing V_β genes of a particular family. As monoclonal antibodies are available against many V_β gene products, this permits enumeration of reactive T cell numbers by techniques essentially analogous to those which allow antigen-reactive B cells to be enumerated. I do not wish to anticipate the chapters to follow except to say that these very powerful techniques have the capacity to reveal much more about the detailed mechanisms of abortion and anergy than our older single cell cloning methods. It is fortunate that the various models to be presented illustrate the full spectrum of alternatives discussed above, not only because such confirmation is reassuring but also because both receptor-transgenic models and those involving superantigens inherently display certain artificial features which require to be checked against "real life." The authors to follow will show that clonal abortion of T cells within the thymus occurs with respect to several self-antigens expressed in the thymus but that peripheral silencing of T cells can also occur, even when the transgene of interest is expressed only in certain cells, such as the β cells of the islets of Langerhans or Schwann cells. Other authors will present data of a persuasive nature suggesting that some kinds of peripheral T cell silencing are due to clonal anergy, while still further models show a true peripheral T cell deletion (as opposed to abortion, by definition confined to the primary lymphoid organs).

The elegant models of B cell tolerance to be presented will likewise show that clonal abortion of B cells in the bone marrow, clonal anergy induction in B cells, and peripheral deletion of exported B cells can all occur. As the chapters in Part II will show, there are still some contentious issues, such as the importance of multivalency of the toleragen and the reasons why some models involving monovalent antigens provoke clonal anergy whereas others suggest no effect at all on the B cell. As with experimental tolerance induction, molar concentration of antigen and affinity of the receptors of a particular B cell for the toleragen will be further important variables. As some degree of autoantibody formation is quite common and consistent with good health, it is clear that purging of the repertoire to remove antiself B cells is incomplete, and the most sensible postulate is of a reciprocal relationship between an antigen's effective concentration and the affinity cutoff point for (functional or actual) clonal elimination. Thus, antigens present at high concentration purge down to low-affinity clonotypes; those present at low concentration can purge only the higher-affinity clonotypes. Effective concentration presumably represents a complex amalgam of molar concentration, epitope valency, and cell surface association. It is a matter of considerable interest that autoantibodies encountered in healthy individuals are usually of the IgM isotype and of low affinity, whereas those present in autoimmune diseases are usually of IgG isotype and higher affinity and frequently (31) but not always (32) bear multiple somatic mutations (33) in their immunoglobulin V genes. It is this issue which we now need to explore in greater detail.

THE GENESIS OF HIGH-AFFINITY ANTIBODIES

Somatic hypermutation of immunoglobulin V genes (33) is an amazing process, with most leaders (34–36) believing that the rate of mutation is about 10^{-3} per base pair per generation, or one mutation every second cell division. Hypermutation is confined to the V region of the fully rearranged gene and takes place at defined times after antigen administration, probably commencing about 5 days after immunization (37) and continuing while B cell memory is being actively laid down. MacLennan and colleagues (38, 39) have provided a coherent theory concerning the cellular physiology of hypermutation. They believe that the process does not occur until after appropriate T cell–dependent antigenic stimulation of a B cell at some location *outside* the germinal center. This first antigenic encounter alters the migratory characteristics of the B cell. Some B cells then find their way to the specialized microenvironment of the germinal center. There, the cells down-regulate expression of the surface immunoglobulin receptor and embark on rapid cell division. It is postulated that hypermutation occurs at this stage. Subsequently, immunoglobulin receptors reappear but cells at this phase of development die through apoptosis unless stimulated by antigen attached to the surface of specialized follicular dendritic cells. The latter, found only in lymphoid follicles, have the capacity to retain antigen long term in an extracellular location (40). Apparently, the CD40 molecule may also be involved in this rescue from B cell death, as anti-immunoglobulin and anti-CD40 antibodies added together to germinal-center-enriched cells *in vitro* provide better protection from death than either alone (39). The postulated reason for this series of events is to arrange a situation where B cells hypermutate their V genes but die unless positively selected by antigen. In the face of gradually appearing serum antibody, the only cells which will have effective access to follicular dendritic-cell-associated antigen will be those with receptors having higher affinity for antigen than the serum antibody, in other words, cells with V gene mutations that have led to affinity maturation.

Given the huge number of self-epitopes (the products of 10^5 genes) and the imprecision in antigen–antibody interactions dictated by the selective nature of immune activation, it is inevitable that B cells will occasionally be driven to autoimmune potential by chance. In other words, a B cell, appropriately responding to some foreign T-dependent antigen, may enter the hypermutation pathway and fortuitously sustain a mutation that confers high affinity to some totally irrelevant self-antigen. It appears likely that rescue from apoptosis is a

T-independent process; at least no cytokine has yet been found which affects it (39). It may be that cross-linking the Ig receptor is the crucial step. So, if the cell were high-affinity antiself to, for example, a membrane antigen of an adjacent cell, its survival and export from the germinal center might be promoted by that self-antigen. Such B cells represent only a veiled threat, as they would have to be activated from the memory state, and T cell tolerance to the antigen concerned would represent a certain safeguard. Still, there are T cell–independent and polyclonal methods of B cell activation, and the very fact that mechanisms for B cell tolerization exist seems to indicate that T cell tolerance alone is not a sufficient bulwark against autoimmunity.

These considerations prompted Linton *et al.* (41) to propose a "second window" of tolerance susceptibility for B cells. We know (42) that B cells maturing from pre-B cells to B cells are especially susceptible to clonal abortion/clonal anergy induction. Might it be that cells which have hypermutated after antigenic stimulation and are now reexpressing their surface receptors pass through a transient stage of great tolerance susceptibility? In that case, encounter with antigen which is *not* follicular dendritic cell associated might cause tolerance. Cells which have mutated toward antiself would encounter self-antigen during this stage and fall down the Lederbergian trapdoor just before reaching follicle-associated antigenic depots.

SOLUBLE ANTIGEN CAN CAUSE ADULT TOLERANCE, INCLUDING A FAILURE OF APPEARANCE OF HIGH-AFFINITY B CELLS

To look at this question, we have recently (43–45) returned to models of adult tolerance involving the injection of freshly deaggregated soluble protein antigens into adult mice. These show that quite small quantities (10 µg or more) of antigen injected before or even somewhat after challenge immunization can cause tolerance. This happens despite maximal T-dependent challenge immunization involving 100 µg of alum-precipitated antigen plus *Bordetella pertussis* bacteria as an adjuvant. Examination of this tolerance phenomenon at the cellular level reveals some interesting features. We have utilized techniques (26,46) that permit high cloning efficiency of murine splenic B cells and, under stimulation with lipopolysaccharide (LPS) and optimal concentrations of IL-4, IL-5, and IL-2, permit 50% of the clonable B cells to switch to IgG_1 production. Clones make 5–20 ng of IgG_1 over 7–9 days of culture. When the supernatants of clonal microcultures are reacted with a protein antigen, such as human serum albumin (HSA), remarkably few splenic B cells from unimmunized mice can make an antibody which, when examined as an IgG_1, detectably binds to HSA. For example, a normal spleen containing 5×10^7 cells or $2.5 \times$

10^7 B cells has only about 100 B cells capable, in this system, of acting as an anti-HSA antibody-forming cell precursor. Moreover, when the enzyme-linked immunosorbent assay (ELISA) system detects one of these rare precursors, the optical density generated and thus presumably the affinity of the antibody are low. After immunization of control mice, this situation does not change detectably for 4 days, but over the next 3 days there is an approximately exponential rise of anti-HSA IgG$_1$ precursors with a doubling period of 9–10 hr. This is consistent with a process requiring some lag (perhaps time for the first extrafollicular encounter with antigen) and then rapid cellular division. Between days 6 and 28, there is a progressive rise in the optical density generated by clonal supernatants. The number of anti-HSA IgG$_1$ AFCP plateaus around day 7 at about 30,000/spleen and gradually declines from about 14 days onward. The preinjection of deaggregated toleragen prior to challenge is accompanied by (1) a marked reduction in clonable anti-HSA IgG$_1$ AFCP noted from day 6 onward such that by day 14 the response is only 1–2% that of controls and (2) a failure of the affinity maturation noted in control clonal cultures.

It is therefore clear that the soluble toleragen has frustrated the processes which lead to the appearance of increased numbers of higher-affinity (presumably V gene mutated) B cells. However, this by no means proves that there has been a direct effect on B cells. It is known that both germinal center formation and isotype switching are largely T cell–dependent phenomena. Therefore the results could be due to the induction of clonal anergy or deletion in T helper cells. Alternatively, they could be due to T cell–mediated suppression. Furthermore, our past experience (e.g., 47) indicates that repertoire purging and suppression can coexist as contributors to one and the same tolerance model. We have therefore constructed an adoptive transfer model for the generation of IgG$_1$-AFCP from mixtures of T and B cell populations injected into lethally X-irradiated mice. The stage that our experiments have reached is that control, transferred populations yield adequate numbers of anti-HSA IgG$_1$-AFCP, and populations from tolerant donors fail to do so. The relevant mixing experiments are in progress.

One of the real barriers to progress in the understanding of B cell memory and of V gene hypermutation is that no one has been able to generate memory or unequivocal mutation *in vitro*. This means that there is a large residual "black box" element to such research, with many of the complex steps occurring only *in vivo* and thus much less amenable to experimental manipulation. It is furthermore a pity that the analysis of CD determinants has progressed so much further in human than in murine systems, whereas T and B cell cloning is more advanced for the mouse. Very recently, attempts to create germinal centers *in vitro* from their component parts have begun in several laboratories, and it is to be hoped that these efforts bear fruit, because so long as studies of B cell activation and tolerization are confined to the germline repertoire, we risk missing an understanding of one of the most fascinating aspects of humoral immunity.

SUMMARY AND CONCLUSIONS

Immunological tolerance revisited in the molecular era looks like replaying the saga of immunoglobulin gene structure and organization, where everyone was right! Repertoire purging, of both T and B cell subsets, is clearly the most important mechanism, although it may not always be deletional, as functional inactivation without cell death also occurs. Probably the most important event is clonal abortion within the primary lymphoid organs, that is, the thymus for T cells and the fetal liver or adult bone marrow for B cells. This will deal with potential high-affinity antiself cells for antigens expressed in the primary lymphoid organs. The capacity to induce anergy or even deletion peripherally by contact between lymphocyte and antigen in the absence of a costimulatory signal remains an important mechanism for inducing tolerance toward antigens expressed only in particular tissues and not within the primary lymphoid organs.

The most unsatisfactory element of tolerance research in the molecular era is that we have not been able to accord the honored place to T cell–mediated suppression which it undoubtedly deserves because we do not have uniformly agreed molecular mechanisms for it. We are becoming more sophisticated about inhibitory effects of lymphokines; this offers one category of solution. The notion that CD8$^+$ T cells can inhibit other T cells without actually killing them is also gaining favor. Clonal methods for enumerating T cell–mediated inhibitory effects require renewed emphasis.

B cells, but not T cells, mutate their receptors' V genes. It is probable that this hypermutation is confined to a particular microenvironment, e.g., the germinal centers, and to particular stages in the life history of the B cell. The question arises whether special mechanisms exist to delete or silence antigenically stimulated B cells which hypermutate to acquire high-affinity antiself activity. This question, initially posed by Klinman's group (41), is now being examined in several laboratories and should be answered soon.

ACKNOWLEDGMENTS

This work was supported by the National Health and Medical Research Council, Canberra, Australia; by grant AI-03958 from the National Institute of Allergy and Infectious Diseases, United States Public Health Service; and by the generosity of a number of private donors to The Walter and Eliza Hall Institute of Medical Research.

REFERENCES

1. Burnet, F. M., and Fenner, F. (1949) *The Production of Antibodies*, 2nd ed., Macmillan, London.
2. Traub, E. (1938) *J. Exp. Med.* **68**, 229–250.
3. Owen, R. D. (1945) *Science* **102**, 400–401.
4. Burnet, F. M., Stone, J. D., and Edney, M. (1950) *Aust. J. Exp. Biol. Med. Sci.* **28**, 291–297.
5. Billingham, R. E., Brent, L., and Medawar, P. B. (1953) *Nature* **172**, 603–606.
6. Jerne, N. K. (1955) *Proc. Natl. Acad. Sci. USA* **41**, 849–857.
7. Talmage, D. W. (1957) *Annu. Rev. Med.* **8**, 239–256.
8. Ehrlich, P. (1900) *Proc. R. Soc. London* **66**, 424–448.
9. Burnet, F. M. (1957) *Aust. J. Sci.* **20**, 67–69.
10. Burnet, F. M. (1959) *The Clonal Selection Theory of Acquired Immunity*, Vanderbilt University Press, Nashville, Tennessee.
11. Jerne, N. K. (1979) *Annu. Rev. Walter and Eliza Hall Inst.* **1978–79**, 34–38.
12. Burnet, F. M. (1956) *Enzyme, Antigen and Virus: A Study of Macromolecular Pattern in Action*, p. 193, Cambridge University Press, Cambridge, UK.
13. Lederberg, J. (1959) *Science* **129**, 1649–1653.
14. Nossal, G. J. V., and Lederberg, J. (1958) *Nature* **181**, 1419.
15. Nossal, G. J. V., and Pike, B. L. (1976) *Immunology* **30**, 189–202.
16. Naor, D., and Sulitzeanu, D. (1967) *Nature* **214**, 687–688.
17. Miller, J. F. A. P. (1961) *Lancet* **2**, 748–749.
18. Good, R. A., Martinez, C., and Gabrielsen, A. E. (1964) *In The Thymus in Immunobiology* (ed. R. A. Good and A. E. Gabrielsen), pp. 3–47, Hoeber, New York.
19. Szenberg, A., and Warner, N. L. (1962) *Nature* **194**, 146–147.
20. Raff, M. C. (1970) *Immunology* **19**, 637–650.
21. Nossal, G. J. V., Pike, B. L., Stocker, J. W., Layton, J. E., and Goding, J. W. (1977) *Cold Spring Harbor Symp. Quant. Biol.* **41**, 237–243.
22. Nossal, G. J. V., and Pike, B. L. (1978) *J. Exp. Med.* **148**, 1161–1170.
23. Metcalf, E. S., and Klinman, N. R. (1976) *J. Exp. Med.* **143**, 1327–1340.
24. Metcalf, E. S., and Klinman, N. R. (1977) *J. Immunol.* **118**, 2111–2116.
25. Nossal, G. J. V., and Pike, B. L. (1980) *Proc. Natl. Acad. Sci. USA* **77**, 1602–1606.
26. Pike, B. L., Alderson, M. R., and Nossal, G. J. V. (1987) *Immunol. Rev.* **99**, 119–152.
27. Pike, B. L., Boyd, A. W., and Nossal, G. J. V. (1982) *Proc. Natl. Acad. Sci. USA* **79**, 2013–2017.
28. Nossal, G. J. V., and Pike, B. L. (1981) *Proc. Natl. Acad. Sci. USA* **78**, 3844–3847.
29. Good, M. F., Pyke, K. W., and Nossal, G. J. V. (1983) *Proc. Natl. Acad. Sci. USA* **80**, 3045–3049.
30. Lamb, J. R., Skidmore, B. J., Green, N., Chiller, J. M., and Feldmann, M. (1983) *J. Exp. Med.* **157**, 1434–1447.
31. Shlomchik, M. J., Marshak-Rothstein, A., Wolfowitz, C. B., Rothstein, T. L., and Weigert, M. G. (1987) *Nature* **328**, 805–811.
32. Dersimonian, H., Schwartz, R. S., Barett, K. J., and Stollar, B. D. (1987) *J. Immunol.* **139**, 2496–2501.
33. Weigert, M. G., Cesari, I. M., Yonkovich, S. J., and Cohn, M. (1970) *Nature* **228**, 1045–1047.
34. McKean, D., Hiippi, K., Bell, M., Staudt, L., Gerhard, W., and Weigert, M. (1984) *Proc. Natl. Acad. Sci. USA* **81**, 3180–3184.
35. Berek, C., and Milstein, C. (1987) *Immunol. Rev.* **96**, 23–42.
36. Rajewsky, K., Förster, I., and Cumano, A. (1987) *Science* **238**, 1088–1094.

37. Levy, N. S., Malpiero, U. V., Lebeque, S. G., and Gearhart, P. J. (1989) *J. Exp. Med.* **169,** 2007–2019.
38. MacLennan, I. C. M., and Gray, D. (1986) *Immunol. Rev.* **91,** 61–85.
39. MacLennan, I. C. M., Liu, Y.-J., Joshua, D. E., and Gray, D. (1989) *Prog. Immunol.* **7,** 443–447.
40. Nossal, G. J. V., and Ada, G. L. (1971) *Antigens, Lymphoid Cells and the Immune Response,* Academic Press, New York.
41. Linton, P.-J., Decker, D. J., and Klinman, N. R. (1989) *Cell* **59,** 1049–1059.
42. Nossal, G. J. V. (1983) *Annu. Rev. Immunol.* **1,** 33–62.
43. Nossal, G. J. V., and Riedel, C. (1989) *Proc. Natl. Acad. Sci. USA* **86,** 4679–4683.
44. Nossal, G. J. V., Karvelas, M., and Lalor, P. A. (1989) *Cold Spring Harbor Symp. Quant. Biol.* **54,** 893–898.
45. Nossal, G. J. V., and Karvelas, M. (1990) *Proc. Natl. Acad. Sci. USA* **87,** 1615–1619.
46. McHeyzer-Williams, M. G. (1989) *Eur. J. Immunol.* **19,** 2025–2030.
47. Good, M. F., and Nossal, G. J. V. (1983) *J. Immunol.* **131,** 2662–2669.

PART II

B CELL TOLERANCE

2

Mechanisms and Meaning of B Lymphocyte Tolerance

DAVID NEMAZEE
Division of Basic Sciences
Department of Pediatrics
National Jewish Center for Immunology and Respiratory Medicine
Denver, Colorado 80206

INTRODUCTION

B lymphocytes have two antigen-specific functions in the immune response: antibody formation and receptor-mediated antigen uptake for presentation to major histocompatibility complex (MHC) class II–restricted T lymphocytes. The outcome of B lymphocyte tolerance leads to two discernible phenotypes, deletion and anergy. What determines these alternative fates and what are the consequences for the immune system?

THE DELETION–ANERGY DECISION

The distinction between these two mechanisms of B cell tolerance is demonstrated *in vivo* either in immunoglobulin transgenic mouse models, in which a large clone of B cells of predefined specificity can be followed *in vivo* in the presence and absence of antigen (1–4), or in certain examples of anti-immunoglobulin M (IgM)–mediated immunosuppression (5–7). In the hen egg lysozyme : anti–hen egg lysozyme (HEL) system, tolerance to the soluble, transgene-encoded, lysozyme leads to functional inactivation, without deletion, of a high-affinity antilysozyme B cell clone, whose IgR are similarly encoded by transgenes (1,2). By contrast, in mice transgenic for immunoglobulin genes encoding an allele-specific anti–MHC class I antibody, autoreactive B cells are eliminated soon after they emerge in the bone marrow, where they first make contact with self-antigen (3,4). Chronic *in vivo* or *in vitro* anti-IgM

19

Molecular Mechanisms of
Immunological Self-Recognition

treatment from early time points in B cell development can lead to profound B cell deletion (5,6) or anergy (7).

To explain the differences in the fates of the autoreactive clones in the various experimental systems, a number of models could be proposed:

1. *The extent of Ig receptor occupancy determines the fate of the B cell.* The extent of B cell surface immunoglobulin receptor occupancy by antigen has been suggested to be an important parameter in B cell tolerance in anti-HEL/HEL double transgenic mice (1) because HEL concentrations leading to 5% receptor occupancy failed to induce B cell tolerance, whereas HEL concentrations resulting in 45% receptor occupancy anergized (2). To explain the deletion with membrane K^k based on a receptor occupancy model one would need to assert that occupancy of even greater than 45% of receptors must be achieved with this membrane antigen to achieve deletion. However, because monovalent ligand is probably unable to transmit signals via immunoglobulin receptor (8), it is likely that the form of HEL that tolerizes in this double transgenic system is aggregated in some way. In addition, for B cell *activation* as few as 10–20 surface Ig molecules need be engaged, provided these are appropriately cross-linked together (9,10). Thus the correlation between the extent of receptor occupancy and tolerance may be misleading, because only a small fraction of the measured antigen may be in a tolerogenic form. Furthermore, Gause *et al.* showed that B cell anergy could be induced *in vivo* in mice with a monoclonal anti-μ antibody under conditions where virtually 100% receptor occupancy was achieved (7). On the other hand, Pike, Boyd, and Nossal showed in *in vitro* experiments that high concentrations of a monoclonal anti-μ could lead to a B cell deletion–like phenomenon, whereas lower doses of the same antibody generated anergic cells (11). This last is consistent with both receptor occupancy and cross-linking models (see item 4 below) because it is possible that the quality of receptor cross-linking is profoundly affected by the antibody/antigen ratio.

2. *The stage of B cell development at which autoantigen is encountered is critical; antigens that are seen in the post–bone marrow stage of B cell development anergize, whereas antigens seen in the bone marrow delete.* This possibility is appealing because of the apparent analogy with T cells: autoreactive T cells encountering class I or class II in the thymus are deleted, whereas those encountering the same antigens in the periphery are anergized. In addition, in a number of systems immature (neonatal spleen or adult bone marrow) B cells require much less antigen for tolerance than more mature, adult splenic B cells (12). However, there is evidence against this simple model for B cells. Bone marrow B cells react with HEL, yet are not deleted (1), but anti–MHC class I B cells that encounter K^k or K^b either in the bone marrow or exclusively in the periphery are deleted (3,4, and our unpublished results).

3. *IgD expression on B cells determines the tolerance phenotype: B cells lacking IgD are deleted, whereas IgD-bearing cells are anergized.* Elements of this idea, albeit in a somewhat different form, have been popular since the discovery that IgD is expressed on most mature B cells but is not expressed on newly generated B cells (13). The available evidence argues against this possibility also: anti-HEL transgenic mice made using IgM + IgD, IgM-only, and IgD-only heavy chain constructs all demonstrated anergy in the presence of lysozyme (14). IgM-only anti-K^k mice delete autoreactive B cells, and IgM + IgD anti-K^k mice have only recently been generated (D. N. Buerki and K. Buerki, unpublished results). It will be of interest if these mice demonstrate anergy rather than deletion.

4. *The extent of surface Ig cross-linking determines B cell fate: strongly cross-linking antigens, like MHC class I, delete, whereas weakly cross-linking antigens, like lysozyme, anergize.* This model has not been adequately tested but, almost by default, appears to best explain the current data. The difficulty here is in defining "strong" and "weak" cross-linking and in explaining in biochemical terms how a quantitative difference in this parameter can determine the qualitative difference in cellular function. Certainly for B cell activation the extent of sIg cross-linking can have profound and striking effects (9).

CONSEQUENCES OF DELETION VERSUS ANERGY

Two concepts are important in discussing the consequences of B cell tolerance: B cell life span and the reversibility of tolerance. It is obvious that deleted B cells have a very short half-life and that the deletion process is, by definition, irreversible. It is less clear whether anergic B cells have a much longer half-life than deleted B cells and whether the anergy is reversible. Most B cells have a lifespan of 1–2 days, but a small subpopulation is admitted to the long-lived, recirculating B lymphocyte pool, which can have a half-life of many weeks (15). Admittance of a B cell to this long-lived pool probably is determined by both its specificity and the extent of competition with other B cells (16). Anergic B cells generated in the HEL : anti-HEL transgenic model appear to have a reasonably long half-life of at least several days in adoptive recipients. It will be of great importance to determine the half-life of these cells in the presence of an excess of competing clones. It is possible that the half-life of the anergic cells under these conditions is rather short, thus minimizing the significance of the functional distinction between the two mechanisms of anergy and deletion, except perhaps in cases of leukopenia. It is of interest in this regard that adults of certain autoimmune-prone mouse strains have profound defects in B cell generation (17,18).

If it proves that anergic B cells indeed have a significant life span, it will be of importance to determine the logic of the immune system in maintaining them. One possibility is that they serve as a pool of B cells that is committed to acquiring V region somatic mutations (19). Another interesting possibility is that the anergic cells, while lacking the ability to produce antibody, retain their antigen-presenting function. These anergic B cells could then have an important suppressor function by specifically taking up and presenting autoantigen in the context of class II antigens and thus providing a specific "sink" for autoreactive T cell help that could otherwise recruit antibody responses from newly emerging, functional autoantigen-specific B cells. Alternatively, the presenting function of anergic B cells may lack the "costimulatory" capacity of normal activated B cells (20) and thus have the capacity to anergize autoreactive T cells directly. The appeal of these ideas is that self-tolerance in the two antigen-specific cells of the humoral response, B and T lymphocytes, would be mutually supportive and synergistic.

REFERENCES

1. Goodnow, C. C., Crosbie, J., Adelstein, S., Lavoie, T. B., Smith-Gill, S. J., Brink, R. A., Pritchard-Briscoe, H., Wotherspoon, J. S., Loblay, R. H., Raphael, K., Trent, R. T., and Basten, A. (1988) *Nature* **334,** 676–682.
2. Goodnow, C. C., Adelstein, S., and Basten, A. (1990) *Science* **248,** 1373–1379.
3. Nemazee, D. A., and Buerki, K. (1989) *Nature* **337,** 562–566.
4. Nemazee, D. A., and Buerki, K. (1989) *Proc. Natl. Acad. Sci. USA* **86,** 8039–8043.
5. Lawton, A. R., and Cooper, M. D. (1974) *Contemp. Top. Immunobiol.* **3,** 193.
6. Cooper, M. D., Kearney, J. F., Gathings, W. E., and Lawton, A. R. (1980) *Immunol. Rev.* **52,** 29–53.
7. Gause, A., Yoshida, N., Kappen, C., and Rajewsky, K. (1987) *Eur. J. Immunol.* **17,** 981–990.
8. Cambier, J. C., Justement, L. B., Newell, M. K., Chen, Z. Z., Harris, L. K., Sandoval, V. M., Klemsz, M. J., and Ransom, J. T. (1987) *Immunol. Rev.* **95,** 37–57.
9. Dintzis, H. M., Dintzis, R. Z., and Vogelstein, B. (1976) *Proc. Natl. Acad. Sci. USA* **73,** 3671–3675.
10. Brunswick, M., June, C. H., Finkelman, F. D., Dintzis, H. M., Inman, J. K., and Mond, J. J. (1989) *Proc. Natl. Acad. Sci. USA* **86,** 6724–6728.
11. Pike, B. L., Boyd, A. W., and Nossal, G. J. V. (1982) *Proc. Natl. Acad. Sci. USA* **79,** 2013–2017.
12. Nossal, G. J. V. (1983) *Annu. Rev. Immunol.* **1,** 33–62.
13. Vitetta, E. S., and Uhr, J. W. (1977) *Immunol. Rev.* 37,50.
14. Basten, A., Brink, R. A., Mason, D. Y. Crosbie, J. and Goodnow, C. C. (1989) *Prog. Immunol.* **7,** 377–384.
15. Fulop, G., Gordon, J., and Osmond, D. G. (1983) *J. Immunol.* **130,** 644–648.
16. Lalor, P. A., Herzenberg, L. A., Adams, S., and Stall, A. M. (1989) *Eur. J. Immunol.* **19,** 507–513.

17. Jyonouchi, H., Kincade, P. W., Landreth, K. S., Lee, G., Good, R. A., and Gershwin, M. E. (1982) *J. Exp. Med.* **155,** 1665–1678.
18. Jyonouchi, H., Kincade, P. W., and Good, R. A. (1985) *J. Immunol.* **134,** 858–864.
19. Linton, P. J., Decker, D., and Klinman, N. (1989) *Cell* **59,** 1049–1059.
20. Mueller, D. L., Jenkins, M. K., and Schwartz, R. H. (1989) *Annu. Rev. Immunol.* **7,** 445–480.

3

Tolerant Autoreactive B Lymphocytes in the Follicular Mantle Zone Compartment: Substrates for Receptor Editing and Reform

CHRISTOPHER C. GOODNOW[1,*], DAVID Y. MASON[2],
MARGARET JONES[2], AND ELIZABETH ADAMS[1]
1. Centenary Institute for Cancer Medicine and Cell Biology
University of Sydney
Sydney NSW 2088, Australia
and
2. Nuffield Department of Pathology
John Radcliffe Hospital
Oxford OX3 9DU, United Kingdom

INTRODUCTION

Humoral immunity is acquired by coordinating genetic events which diversify B lymphocyte antigen receptors with cellular mechanisms for positively or negatively selecting B cells with particular specificities. Selection of an appropriate repertoire of B cells appears to involve a rather complex series of cellular events, as suggested by the numerous distinct developmental and functional phenotypes exhibited within the B cell lineage (see ref. 1); by the complicated differential expression of immunoglobulin M (IgM), IgD, and downstream isotypes of antigen receptor (2); and by the discrete and specialized microenvironments in which different B cell subpopulations are located (3). How these various subpopulations are related to one another and to the selection

°Current address: Howard Hughes Medical Institute and Department of Microbiology and Immunology, Beckman Center, Stanford University School of Medicine, Stanford, CA 94305.

Molecular Mechanisms of
Immunological Self-Recognition

processes which ensure B cell memory and self-tolerance nevertheless remains obscure, due mainly to the technical difficulty in following the development and fate of B cells expressing a particular antibody specificity among the large numbers of cells with unrelated specificities. These problems have been particularly acute in the case of tolerance, where self-reactive B cells have long been thought to be either physically eliminated (4) or functionally inactivated (5).

Mice carrying rearranged immunoglobulin transgenes, in which large numbers of B cells express transgene-encoded immunoglobulin with a single, defined antigen specificity (6), offer a unique opportunity to study B cell selection because the development and fate of these cells can easily be tracked *in vivo*. To focus on mechanisms of self-tolerance, we have produced transgenic mice carrying rearranged Ig genes encoding both IgM and IgD immunoglobulins (carrying the complete μ-δ heavy chain locus) with high affinity for hen egg lysozyme (HEL), in which many of the B cells express only the transgene-encoded antilysozyme receptors (7). By mating these Ig-transgenic animals with transgenic mice expressing hen egg lysozyme itself, "double-transgenic" progeny were generated in which the antilysozyme B cells were rendered functionally tolerant (anergic) but not physically eliminated (7). In this chapter, we first review the developmental pattern of events accompanying acquisition of B cell tolerance in these animals and discuss why these events may differ from those observed in anti–major histocompatibility complex (MHC) Ig-transgenic mice (8, 9). In addition, we speculate on the physiological role which might be played by tolerant B cells within the follicular mantle zone and the possibility that progeny from these cells may be "reformed" by undergoing hypermutation and further selection within adjacent germinal centers.

LOCALIZATION OF TRANSGENE-EXPRESSING B CELLS WITHIN PERIPHERAL LYMPHOID ORGANS

As detected by flow cytometry of lymphoid cell suspensions, 80–90% of the B cells in the antilysozyme Ig-transgenic mice and double-transgenic littermates express transgene-encoded (a-allotype) IgM and IgD, although IgM is selectively modulated on the double-transgenic B cells (Fig. 1). Immunohistological techniques were used to determine the anatomical distribution of B cells expressing transgene-encoded antigen receptors in antilysozyme IgM/IgD-transgenic mice and lysozyme-expressing double-transgenic littermates (10) (summarized in Fig. 2). In the Ig-transgenic mice, labeling of spleen sections using monoclonal antibodies specific for IgM[a] (transgene allotype) or IgM[b] (endogenous allotype), revealed well-developed B cell areas within the white pulp and confirmed that most of the B cells in the follicular mantle zone and in the splenic marginal zones expressed only transgene-encoded heavy chains. Ex-

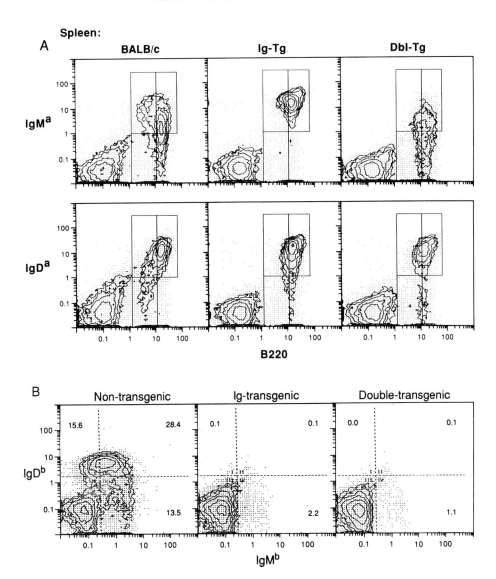

Fig. 1. Flow cytometric analysis of Ig expression in Ig- and double-transgenic mice. (A) FACS analysis of spleen cells from a nontransgenic BALB/c mouse (IgHᵃ strain), and from Ig-transgenic and double-transgenic C57BL/6 mice (IgHᵇ strain), illustrating expression of transgene-encoded IgM (IgMᵃ) and IgD (IgDᵃ) on the majority of B220⁺ B cells. (B) FACS analysis of spleen cells from a nontransgenic C57BL/6 mouse and from littermate Ig- and double-transgenic mice, stained for endogenous IgM (IgMᵇ) and IgD (IgDᵇ). Note that the few b-allotype-expressing cells present in the transgenic mice are predominantly IgM^high and IgD^low.

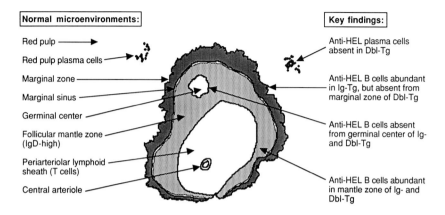

Fig. 2. Summary of histological findings in lysozyme/antilysozyme Ig-transgenic (Ig-Tg) and double-transgenic (Dbl-Tg) mice. The diagram illustrates the normal cross-sectional structure of a splenic white pulp cord in the mouse, with the predominant microenvironments labeled on the left. On the right-hand side, the key observations in the transgenic mice are summarized.

pression of both H and L transgene products on peripheral B cells was confirmed by staining spleen and lymph node sections for lysozyme-binding activity, since high-affinity lysozyme binding is dependent on pairing of the transgene-encoded heavy chain with the correct light chain. Staining for IgM and IgD demonstrated that the B cells in the follicular mantle zones of spleen and lymph node expressed both IgM and IgD, whereas those in the splenic marginal zones expressed IgM but little or no IgD, indicating normal differential regulation of transgene-encoded IgM and IgD between these two anatomical subpopulations (11, 12).

By contrast with the abundance of transgene-expressing B cells in the follicular mantle zone and marginal zone microenvironments, no such cells were detected within the germinal centers of Ig-transgenic animals, and relatively few were present in the red pulp as plasma cells. Instead, endogenous-(IgM^b) expressing cells appeared to account for the majority of cells in both of these compartments. By fluorescence-activated cell sorter (FACS) analysis, the endogenous-expressing B cells were predominantly IgM^{high} and IgD^{low} (Fig. 1B), which would be consistent with the phenotype of germinal center cells, marginal zone cells, or Ly1 B cells.

Comparison of Ig-transgenic mice with tolerant double-transgenic littermates revealed a marked change in the distribution of transgene-expressing B cells (10) (summarized in Fig. 2). Thus, lysozyme-binding B cells were absent from the splenic marginal zones and restricted to the follicular mantle zones in double-transgenic mice, the pattern of staining coinciding with that of IgD-bearing cells. The marginal zones were thinner overall, although they still contained a small number of residual B cells expressing high levels of IgM. Cells expressing IgM^b accounted, at least in some mice, for many of the residual nonlysozyme binding marginal zone B cells. Lysozyme-binding plasma cells were rare in the red pulp of double-transgenic animals, in keeping with the lack of plaque-forming cells (7). In all other respects, the overall lymphoid architecture was not noticeably different in the double-transgenic mice compared with the Ig-transgenic controls.

TIMING OF TOLERANCE INDUCTION DURING BONE MARROW B CELL DIFFERENTIATION

The presence of lysozyme-reactive B cells in the follicles of the lysozyme double-transgenic mice could in theory have been due to failure of the B cells to encounter lysozyme earlier in their development within the bone marrow. To follow earlier events in the Ig- and double-transgenic mice, differentiating B lineage cells in the bone marrow were analyzed flow cytometrically by measuring expression of surface Ig and the B lineage–specific cell surface antigen B220 (13–15). Subsets of differentiating B lineage cells which were phenotypically equivalent to pro-B/pre-B blasts, small pre-B cells, and immature B cells were present in the bone marrow of IgM/IgD Ig-transgenic mice (Fig. 3). The total number of immature B220^{low} cells was markedly reduced in the Ig-transgenic mice (10), as has been noted in other Ig-transgenic animals (16). Most of the reduction reflected diminished numbers of surface Ig-negative pro-B or pre-B cells, consistent with the evidence that expression of a productively rearranged heavy chain gene directly triggers exit from the proliferative phase of pre–B cell development (17, 18). Like immature B cells in nontransgenic animals (2, 15), newly differentiated sIg$^+$ B220^{low} B cells in the bone marrow of Ig-transgenic mice expressed transgene-encoded IgM but lacked cell surface expression of IgD, whereas mature B220^{high} B cells bore high levels of both isotypes (Fig. 3). Thus, differential RNA splicing of the rearranged VDJ$_H$ exon to either the μ- or δ-constant region exons (2) was developmentally regulated in a normal fashion within the introduced transgenic locus, and the B cells progressed normally through an immature IgM-only phase of development.

To determine if lysozyme was encountered by immature B cells in the bone marrow of the lysozyme double-transgenic mice, receptor-bound lysozyme was detected using a different antilysozyme monoclonal antibody, HyHEL5 (19). Using this assay, levels of receptor-bound endogenous lysozyme comparable to those observed in the periphery could be demonstrated on all of the immature B220^{low} B cells in the double-transgenic mice (Fig. 3). In addition to receptor occupancy, these newly differentiated B cells exhibited a 2- to 5-fold decrease

Bone marrow:

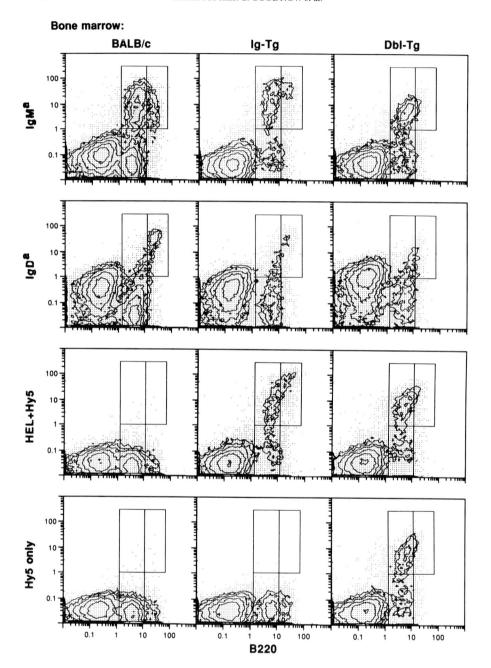

Fig. 3. Differential expression of transgene-encoded IgM and IgD antigen receptors and binding of lysozyme during B cell development. FACS analysis of bone marrow cells from MD3×ML5

in cell surface IgM relative to the IgM levels on B220^{low} cells in Ig-transgenic littermates. The decreased expression of membrane IgM on immature B cells was similar, although of lesser degree, to the 10- to 20-fold down-regulation of membrane IgM on mature peripheral B cells in the double-transgenic mice (7, 20). Adoptive transfer experiments confirmed that the immature B cells had indeed been rendered functionally tolerant by this early encounter with self-lysozyme (10).

RECOVERY FROM RECEPTOR MODULATION AND TOLERANCE

We previously found that tolerance in the antilysozyme B cells was accompanied by modulation of membrane IgM expression (7), the two events appearing closely correlated in dose–response studies *in vivo* (20). Recent studies indicate that this receptor modulation is reversible. First, when the double-transgenic B cells are removed from their tolerizing environment and "parked" *in vivo* in nontransgenic hosts, they regain control levels of membrane IgM with a slow tempo, with partial recovery evident after 1–2 days (Fig. 4A) and full recovery by 7–10 days (21). By contrast, the same B cells parked in a lysozyme-expressing recipient animal show no recovery of their IgM^{low} phenotype. The modulated IgM^{low} phenotype is also rapidly acquired by nontolerant Ig-transgenic B cells parked in lysozyme-expressing recipients (Fig. 4A). Second, tolerant double-transgenic B cells recover IgM expression with a rapid tempo *in vitro* when stimulated with lipopolysaccharide (LPS), full recovery under these conditions being attained within 36 hours (Fig. 4B). As with the *in vivo* studies, the presence of lysozyme in the culture medium prevents recovery of receptor expression in the double-transgenic B cells and induces rapid modulation of membrane IgM on nontolerant Ig-transgenic controls (Fig. 4C).

Tolerant B cells from the double-transgenic mice can also make a functional recovery when removed from constant exposure to autologous lysozyme (21), although the tempo and conditions under which recovery occurs suggest that

---

Ig-transgenic or double-transgenic littermates and from an age-matched nontransgenic BALB/c control, stained for B220 and for IgM^a (*top row*) or IgD^a (*second row*), demonstrating selective expression of IgM but not IgD on immature B220^{low} cells. Windows on B220^{low} and B220^{high} cells are shown. To detect lysozyme bound to receptors on developing B cells, the same suspension of bone marrow cells was stained (*third row*) with saturating concentrations of lysozyme followed by biotinylated antilysozyme antibody, HyHEL5, to reveal all cells with the potential to bind lysozyme. Alternatively, (*bottom row*) the cells were stained with HyHEL5–biotin alone, revealing only cells with *in vivo*–derived lysozyme already bound to their receptors. Equivalent analyses for the MD4 Ig-transgenic line are shown in ref. 10.

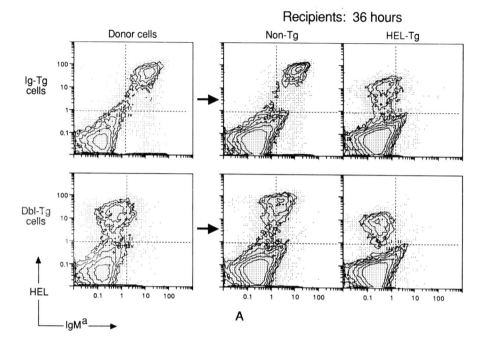

Fig. 4.   Recovery of membrane IgM expression on tolerant B cells following removal from constant exposure to lysozyme. (A) FACS analysis of IgM expression on splenic B cells from Ig- or double-transgenic mice, after transfer into irradiated nontransgenic (Non-Tg) or lysozyme-transgenic (HEL-Tg) mice. Partial recovery of membrane IgM expression on the tolerant Dbl-Tg B cells is already evident in the nontransgenic recipient, compared with either the starting donor cells or cells returned to a lysozyme-containing recipient. Conversely, rapid modulation of membrane IgM is induced in nontolerant Ig-Tg cells after transfer into lysozyme-expressing recipients. (B) Recovery of IgM expression on tolerant double-transgenic splenic B cells after short-term culture *in vitro*. Spleen cells were cultured for 2 days with no stimuli (medium only) or with the B cell mitogens anti-IgD dextran (anti-IgD-dex) and lipopolysaccharide (LPS), either in the absence of lysozyme (*top row*) or with 100 ng/ml lysozyme also present in the culture medium (+ HEL, *bottom row*). The cells were compared with the freshly isolated starting cells (Day O) by FACS analysis. (C) FACS analysis of cultures parallel to those in (B) but using nontolerant Ig-transgenic spleen cells. (*Figure continues.*)

additional events contribute to the functionally silenced state. For example, tolerant double-transgenic B cells which have fully recovered their receptor expression after being parked for 10 days in nontransgenic mice make only a low-level antibody response after primary stimulation with an immunogenic lysozyme conjugate, but mount a good response to a second stimulation. The poor antibody responses observed initially *in vivo* and with LPS stimulation *in vitro* (21) suggest that the tolerant B cells are reversibly blocked in their capacity to differentiate into plasma cells.

Dbl-Tg cells:

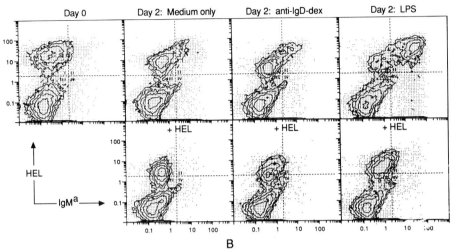

B

Ig-Tg cells:

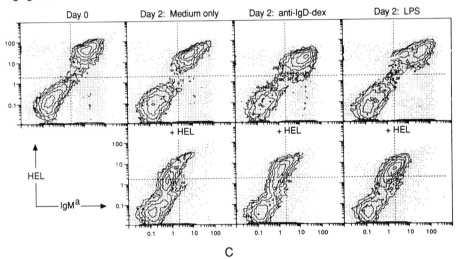

C

Fig. 4. (*continued*)

DELETION OR ANERGY?

The findings reviewed here indicate that maturation and emigration of B cells from the bone marrow to the follicular mantle zone occurred in the double-transgenic animals after the cells had become functionally tolerant, since lysozyme binding and tolerance induction were concomitant with surface IgM expression in the marrow. It is unlikely that the maturation observed arises simply through a minority of cells escaping tolerance induction at the immature stage, since equivalent numbers of tolerant and nontolerant mature B cells are observed in bone marrow radiation chimera studies even at early stages of reconstitution. This continued maturation of self-reactive B cells contrasts with results obtained in anti-MHC Ig-transgenic mice (8, 9) and in our recent studies of double-transgenic mice expressing a membrane-bound form of lysozyme (22), in which recognition of highly multivalent self-antigens by immature B cells completely aborted further maturation of the cells. The fact that binding of high levels of soluble lysozyme, which is unlikely to cross-link antigen receptors efficiently, failed to abort maturation but nevertheless triggered a state of functional unresponsiveness in the B cells indicates that induction of deletion is not simply a consequence of engaging antigen receptors on immature B cells. Rather, the contrasting responses to soluble and membrane-bound antigens in immature B cells suggest that membrane Ig signaling is not a simple all-or-none process and that functionally distinct signaling thresholds may be differentially activated depending on the degree of receptor cross-linking.

DOES THE FOLLICULAR MANTLE ZONE SERVE AS A "REFORM SCHOOL" FOR WAYWARD B CELLS?

The selective accumulation of tolerant lysozyme-reactive B cells in the follicular mantle zones raises the questions of whether significant numbers of self-reactive B cells exist in a similar state in normal polyclonal repertoires and what possible function such cells might serve. Under normal conditions, IgD[high] follicular mantle zone B cells are predominantly long-lived and recirculate between follicles via the lymph and blood (15, 23, 24). Although a selected set of V genes appears to be expressed by these cells, they differ from the predominant population of antigen-expanded memory B cells in that they show little or no V gene hypermutation (25). IgD[high] B cells normally exhibit a broad range of membrane IgM expression (26) and, as pointed out previously (20), a significant fraction display an IgD[high], IgM[low] phenotype comparable to that of the tolerant self-reactive B cells located in this site in the double-transgenic mice.

One possible physiological rationale for the persistence of moderately self-reactive B cells within the follicular mantle zone is that they may help to mini-

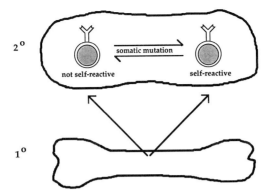

Fig. 5. Hypothetical two-way flow of cells between autoreactive and nonautoreactive pools as a result of Ig gene hypermutation in peripheral (secondary, 2°) lymphoid organs. As discussed in the text, silenced autoreactive B cells persisting in the follicular mantle zone may help to "fill in" holes in the antibody repertoire by generating nonautoreactive variant cells through germinal center hypermutation.

mize the occurrence of holes in the B cell repertoire, as might occur if all self-reactive B cells were eliminated. It is now apparent that considerable Ig diversity is generated in B cells following antigenic stimulation, by Ig gene hypermutation within germinal center reactions (27, 28). The persistence of the tolerant double-transgenic B cells within the follicular mantle zone leaves them in a strategic location with respect to the hypermutation process, since germinal center reactions form in the midst of the follicular mantle zone. Moreover, although the silenced B cells are inhibited with respect to antibody secretion, they are still capable of efficient proliferation in response to LPS (21) and possibly to helper T cells. It is therefore conceivable that silenced self-reactive B cells in the follicular mantle zone might be recruited into the germinal center reaction on the basis of some affinity for a foreign antigen and undergo activation and Ig gene hypermutation. Through substitutions in the receptor binding site, useful daughter cells might be expected to arise which retained antiforeign specificity but had lost all autoreactivity (Fig. 5).

REFERENCES

1. Moller, G., ed. (1986) *Immunol. Rev.* **93**.
2. Blattner, F., and Tucker, P. W. (1984) *Nature* **307,** 417.
3. MacLennan, I. C. M., Oldfield, S., Liu, Y.-J., and Lene, P. J. L. (1989) *Curr. Top. Pathol.* **79,** 37.
4. Burnet, F. M. (1959) *The Clonal Selection Theory of Acquired Immunity,* Vanderbilt University Press, Nashville.

5. Nossal, G. J. V. (1983) *Annu. Rev. Immunol.* **1,** 33.
6. Storb, U. (1987) *Annu. Rev. Immunol.* **5,** 151.
7. Goodnow, C. C., Crosbie, J., Adelstein, S., Lavoie, T. B., Smith-Fill, S. J., Brink, R. A., Pritchard-Briscoe, H., Wotherspoon, J. S., Loblay, R. H., Raphael, K., Trent, R. J., and Basten, A. (1988) *Nature* **334,** 676.
8. Nemazee, D. A., and Bürki, K. (1989a) *Nature* **337,** 562.
9. Nemazee, D., and Bürki, K. (1989b) *Proc. Natl. Acad. Sci. USA* **8,** 8039.
10. Mason, D. Y., Jones, M., and Goodnow, C. C. (1992) *Int. Immunol.* **4,** 163–175.
11. Stein, H., Bonk, A., Tolksdorf, G., Lennert, K., Rodt, H., and Gerdes, J. (1980) *J. Histochem. Cytochem.* **28,** 746.
12. Gray, D., MacLennan, I. C. M., Bazin, H., and Khan, M. (1982) *Eur. J. Immunol.* **12,** 564.
13. Kincade, P. W. (1981) *Adv. Immunol.* **31,** 177.
14. Coffman, R. L., and Weissman, I. L. (1983) *J. Mol. Cell. Immunol.* **1,** 31.
15. Förster, I., Vieira, P., and Rajewsky, K. (1989) *Int. Immunol.* **1,** 321.
16. Herzenberg, L. A., Stall, A. M., Braun, J., Weaver, D., Baltimore, D., Herzenberg, L. A., and Grosschedl, R. (1987) *Nature* **329,** 71.
17. Reth, M., Petrac, E., Wiese, P., Lobel, L., and Alt, F. W. (1987) *EMBO J.* **6,** 3299.
18. Kitamura, D., Roes, J., Kühn, R., and Rajewsky, K. (1991) *Nature* **350,** 423.
19. Smith-Gill, S. J., Mainhart, C. R., Lavoie, T. B., Rudikoff, S., and Potter, M. (1984) *J. Immunol.* **132,** 963.
20. Goodnow, C. C., Crosbie, J., Jorgensen, H., Brink, R. A., and Basten, A. (1989b) *Nature* **342,** 385.
21. Goodnow, C. C., Adams, E., and Brink, R. A. (1991) *Nature* **352,** 532.
22. Hartley, S. B., Crosbie, J., Brink, R., Kantor, A. B., Basten A., and Goodnow, C. C. (1991) *Nature* **353,** 765.
23. MacLennan, I. C. M., and Gray, D. (1986) *Immunol. Rev.* **91,** 61.
24. Sprent, J., and Basten, A. (1973) *Cell. Immunol.* **7,** 40.
25. Gu, H., Tarlinton, D., Müller, W., Rajewsky, K., and Förster, I. (1991) *J. Exp. Med.* **173,** 1357.
26. Hardy, R. R., Hayakawa, K., Parks, D. R., and Herzenberg, L. A. (1983) *Nature* **306,** 270.
27. Rajewsky, K., Forster, I., and Cumano, A. (1987) *Science* **238,** 1088.
28. Berek, C., and Milstein, C. (1988) *Immunol. Rev.* **105,** 5.

PART III

LYMPHOCYTE SIGNALING

4

T Cell Anergy

BART BEVERLY, LYNDA CHIODETTI, KURT A. BRORSON,
RONALD H. SCHWARTZ, AND DANIEL MUELLER*
Laboratory of Cellular and Molecular Immunology
National Institute of Allergy and Infectious Diseases
National Institutes of Health
Bethesda, Maryland 20892

INTRODUCTION

During T cell development in the thymus, self-reactive cells are eliminated by a process known as negative selection. This mechanism for maintaining self-tolerance is, however, effective only for the self-antigens (self-Ag) present in the thymus; therefore, the immune system must have a way of dealing with self-Ag present only in the periphery. One such mechanism that has been proposed is the functional inactivation of self-reactive T cells, a state known as clonal anergy. This state is thought to be induced when T cells are stimulated with T cell receptor (TCR) occupancy, signal 1, in the absence of a biochemically distinct signal provided by antigen-presenting cells (APC), signal 2 (1–3).

Several methods of inducing a state of T cell clonal anergy *in vitro* in mouse T cell clones of the type 1 (T_H1) phenotype have been reported. These models of anergy have permitted the study of the cellular, biochemical, and molecular characteristics of anergic cells, and the results of such experiments are presented in this chapter.

CELLULAR CHARACTERISTICS

Stimulation of anergic T_H1 cells with Ag and APC results in a diminished proliferative response compared to that of resting controls (4–9) (see Fig. 1A). There is also a decrease in the amount of lymphokines produced by these cells on reactivation. Figure 1B through 1E compare the amounts of

Current address: Department of Medicine, University of Minnesota Hospitals, Box 108, UMHC, Minneapolis, MN 55455

Molecular Mechanisms of
Immunological Self-Recognition

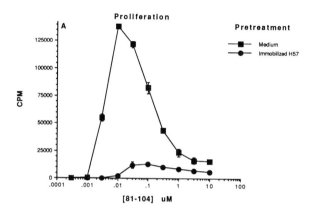

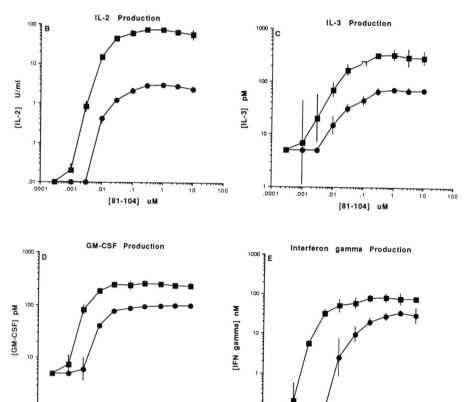

the lymphokines interleukin-2 (IL-2), interleukin-3 (IL-3), granulocyte-macrophage colony-stimulating factor (GM-CSF), and interferon-γ (IFN-γ), respectively, produced on activation of anergized vs. normal cells. The decrease in IL-2, the autocrine growth factor, is the cause of the decreased proliferation observed in anergized cells. The decrease in lymphokine production is graded, with IL-2 > IL-3 > GM-CSF and IFN-γ (compare Fig. 1B–E). This decrease in lymphokine production is correlated with a decrease in the steady-state mRNA levels (Fig. 2A–D). In other in vitro systems, there is evidence that IL-3 (12) and IFN-γ (13) production by lymphocytes is dependent on IL-2. Whether the observed decrease in the production of IL-3, GM-CSF, and IFN-γ by anergic cells is due, in part or entirely, to the IL-2 production defect is currently unknown. These results indicate that the major defect in anergic cells is their ability to produce IL-2 in response to Ag and APC.

The IL-2 production defect is not simply due to the fact that the cells have been recently activated (preactivation). The ability of resting cells to produce IL-2 was compared for cells previously stimulated with concanavalin A (Con A) alone (anergic), Con A plus APC (complete preactivation), or IL-2 (lymphokine preactivation). Figure 3 shows that preactivated cells are capable of producing as much as or more IL-2 than the unstimulated controls. The only noticeable effect is a shift in the dose–response curve such that higher concentrations of Ag are required for half-maximal IL-2 production. The anergic cells show this

---

Fig. 1. Functional characteristics of anergic T cells. Viable A.E7, a normal mouse T_H1 clone reactive to carboxyl-terminal residues 81 to 104 of pigeon cytochrome c and IE^k (2×10^4), pretreated with (filled circles) or without (filled squares) immobilized anti-TCR antibody (H57, 10 μg/ml) 10 days earlier, were stimulated in triplicate with the indicated concentrations of Ag and 5×10^5 T-depleted 3000-rad-irradiated B10.A splenocytes in a final volume of 200 μl in microtiter wells. Twenty-four hours later, 150 μl of supernatant was removed for analysis of lymphokines, and each well was pulsed with [³H]thymidine and harvested 18 hours later onto glass fiber filters. Incorporated radioactivity was determined by liquid scintillation counting. (A) Proliferation of H57-pretreated (anergic) vs. medium-pretreated (control) cells. The results are expressed as the average cpm ± SD from triplicate cultures. (B) IL-2 production from anergized vs. control cells. IL-2 was determined using the IL-2–dependent line CTLL-20 obtained from American Type Culture Collection (Rockville, MD). The results are expressed as the geometric mean ×/÷ SD obtained from triplicate determinations; the limit of detection of the assay was 0.01 units/ml, a unit being (dilution)⁻¹ giving half-maximal proliferation of CTLL-20. Samples scoring negative are presented as 0.01 units/ml. (C, D, and E) Lymphokine production (IL-3, GM-CSF, and IFN-γ, respectively) from anergized vs. control cells. The lymphokines were quantitated using an enzyme-linked immunosorbent assay (ELISA) with recombinant lymphokines as standards. The results are expressed as the geometric mean ×/÷ SD obtained from triplicate determinations; the limit of sensitivity was 5 pM for IL-3 and GM-CSF and 0.1 nM for IFN-γ. Samples scoring negative are presented as 5 pM or 0.1 nM, respectively. No lymphokines or proliferation was detected in samples stimulated in the absence of Ag.

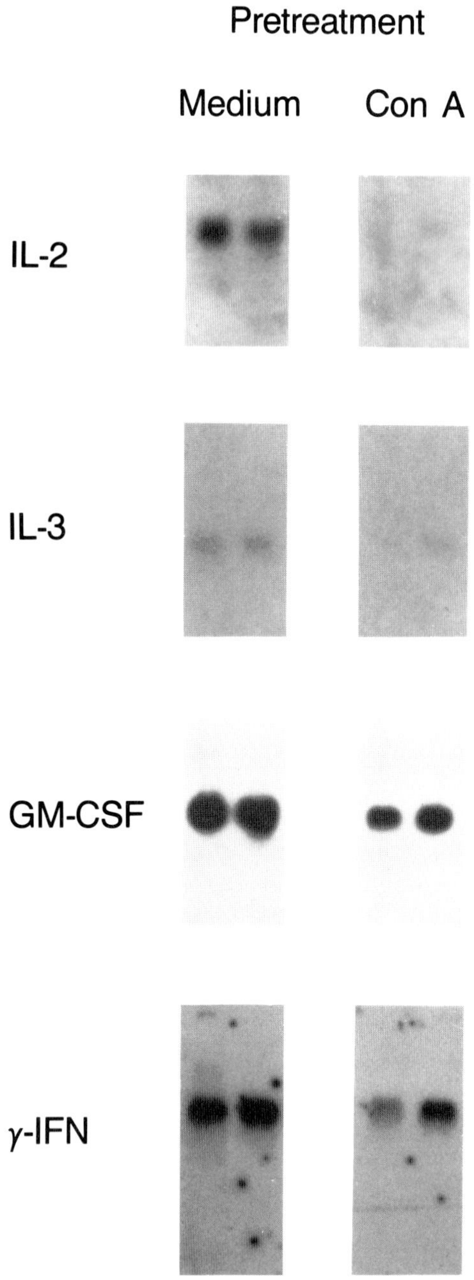
Pretreatment
Medium
Con A
IL-2
IL-3
GM-CSF
γ-IFN

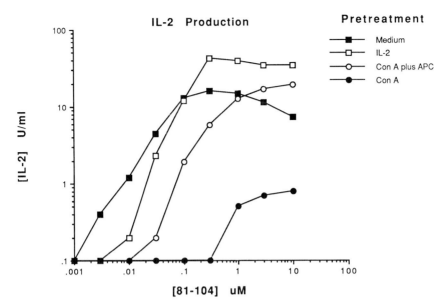

Fig. 3. Comparison of IL-2 production by resting, prestimulated, and anergized T cells. Viable A.E7 cells (2×10^4) pretreated 8 days earlier with medium (*filled squares*), Con A plus APC (*open circles*), IL-2 from supernatant of the gibbon leukemic T cell line MLA 144 (*open squares*), or Con A (*filled circles*) were stimulated with Ag and APC (as described in the legend to Fig. 1) in duplicate cultures. At 24 hours, 100 μl of supernatant was removed and the duplicate samples were pooled and analyzed for IL-2 as described in the legend to Fig. 1. The limit of detection of the assay was 0.1 U/ml, and samples scoring negative are presented as 0.1 U/ml. No IL-2 was detected in any group stimulated in the absence of Ag.

Fig. 2. Northern blot analysis of lymphokine mRNA. Levels of IL-2, IL-3, GM-CSF, and IFN-γ steady-state mRNA are lower in anergic A.E7 T cells after restimulation by antigen and APC than in restimulated normal A.E7 T cells. A.E7 T cells were anergized by stimulation with 5 μg/ml Con A for 24 hours followed by addition of 10 mg/ml α-methylmannoside and rest for 5 days. Normal T cells were coincubated without the Con A but in the presence of α-methylmannoside. Total mRNA was prepared in duplicate from each group by ultracentrifugation over a CsCl cushion, of which 10 μg was run per lane in a 1.0% agarose formaldehyde denaturing gel and transferred to Biotrans membranes (ICN, Irvine, CA). Two separate blots from independent experiments were probed with mouse lymphokine cDNA probes. The first was hybridized with an IL-2 probe (obtained from ATTC, Rockville, MD) and a GM-CSF probe (kindly donated by N. Gough, Melbourne, Australia; ref. 10). The second was hybridized with an IL-3 probe (obtained from ATTC) and an IFN-γ probe (obtained from K. Arai, Palo Alto, CA; ref. 11). The hybridized blots were washed twice at high stringency ($0.1 \times$ SSC, 60°C) and exposed to autoradiography for 3 hours to 10 days.

effect in a more pronounced way and in addition demonstrate a decrease in the maximal IL-2 production not seen in other preactivated groups.

Anergic cells are not moribund, as evidenced by the fact that they can be stimulated to divide by addition of exogenous IL-2 (4–8). They also can be stimulated with the calcium ionophore ionomycin plus the phorbol ester phorbol myristate acetate (PMA) (5). This pharmacologic stimulus is capable of mimicking some of the effects of TCR signals but may go beyond this because of the irreversible nature of the protein kinase C (PKC) activation.

One possible explanation for the decreased IL-2 production could be the decreased surface expression of molecules important in T cell activation. Three such molecules are the α-β chains of the TCR, CD3, and CD2; their expression on anergized and resting cells is shown in Fig. 4. The level of TCR and CD3 is only slightly lower on the anergized cells than on the controls (compare the dashed and solid lines, with the mean fluorescence intensities of 50 *vs.* 66 and 14 *vs.* 19, respectively). The expression of CD2 is slightly higher on anergized cells than normals (mean fluorescence intensities of 56 *vs.* 46). These results, which are consistent with previous findings for TCR (5) and CD3 (8), are evidence that the IL-2 production defect is not the result of down-modulation of these cell surface molecules.

BIOCHEMICAL EVENTS RESULTING FROM TCR OCCUPANCY

One early result of TCR occupancy is the activation of a phospholipase C which gives rise to two second messengers inositol 1,4,5-trisphosphate (IP_3) and diacylglycerol. IP_3 leads to an increase in intracellular calcium, and diacylglycerol leads to the translocation and subsequent activation of PKC (reviewed in ref. 14). The ability of anergic cells to mediate activation of phospholipase C and the resulting second messengers after TCR occupancy has been examined in TH1 clones. The increase in water-soluble inositol phosphates in normal and anergized cells is shown in Fig. 5. As can be seen, the cells are anergic as measured by their inability to respond to Ag and APC (compare open bars to hatched bars in Fig. 5A). Figure 5B shows that anergized cells generate 77% as much water-soluble inositol phosphates as controls in response to Ag plus APC and 74% as much in response to aluminum fluoride, a nonspecific activator of G proteins and phospholipase C. These results indicate that the generation of inositol phosphates by anergic cells following activation through the TCR is essentially normal.

The initial water-soluble inositol phosphate generated is inositol 1,4,5-trisphosphate, which is responsible for a rise in the concentration of intracellular calcium, $[Ca^{2+}]_i$. The ability of resting, preactivated, or anergic cells (see Fig.

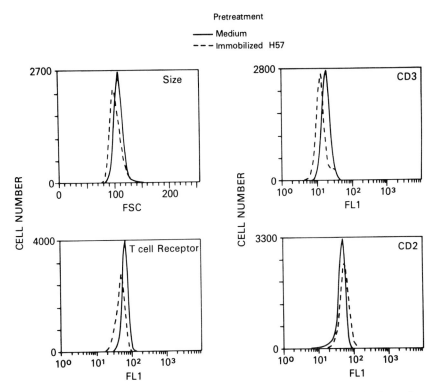

Fig. 4. Surface expression of TCR, CD3, and CD2 on normal and anergized cells. Medium (*solid line*) or anergic (*dashed line*) A.E7 cells (5×10^5) pretreated as described in the legend to Fig. 1 were stained with antibodies against the $\alpha\beta$ chains of the TCR (TCR), ε chain of the CD3 complex (CD3), or CD2 (CD2) and analyzed on a FACScan; 50,000 events gated on live cells were collected using FITC-conjugated H57, biotinylated 500.A2, or biotinylated R2-5, respectively, followed by FITC-avidin.

6A) to increase $[Ca^{2+}]_i$ in response to two different doses of Con A is shown in Fig. 6B. As can be seen, the ability of anergic cells to increase $[Ca^{2+}]_i$ is not significantly different from that of preactivated or resting controls.

The appearance of free inositol phosphates and the increase in calcium ions in the cytoplasm of anergic T cells soon after stimulation with Con A, anti-CD3 monoclonal antibody (mAb), or Ag/APC suggested that CD3-mediated proximal signaling pathways remained intact. Subtle shifts in stimulus dose–response curves were frequently seen in these experiments; however, the minor desensitization never correlated with the magnitude of the proliferative defect. Furthermore, similar changes in sensitivity were also typically observed in T cells preactivated with normal APC, suggesting that the

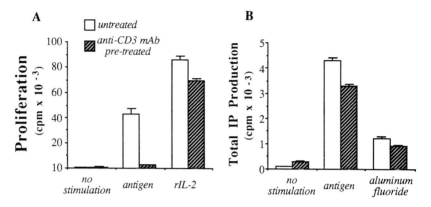

Fig. 5. Proliferation and inositol phosphate generation of resting and anergized T cells. Resting A.E7 cells were cultured in 6-well Costar plates coated with (*hatched bars*) or without (*open bars*) 1 μg/ml of the antimouse CD3 antibody 145-2C11. Twenty-four hours later, the cells in 2C11-coated wells were transferred to uncoated plates. Six days later, both groups were harvested and dead cells and debris removed by centrifugation over Ficoll Hypaque gradients. (A) Proliferation was measured by [³H]thymidine incorporation for 18 hours following a 48-hour stimulation of the cells with 1.0 μM DASP, a synthetic analog of pigeon cytochrome *c* residues 81–104, plus irradiated APC (antigen) or 200 U/ml human recombinant IL-2 (rIL-2). (B) Inositol phosphate generation following stimulation with Ag plus APC. A.E7 cells were loaded with myo-[³H]inositol overnight. The labeled cells were washed, treated with Ficoll, and stimulated in the presence of 10 mM lithium chloride with 100 μM DASP plus APC (antigen) or 10 μM aluminum chloride and 10 μM sodium fluoride (aluminum fluoride). After 1 hour of activation, the water-soluble inositol phosphates were determined as previously described (7).

desensitization was associated with recent stimulation rather than the development of anergy *per se*.

Additional experiments examined the consequences of phosphatidylinositol bisphosphate hydrolysis in the anergic T cells. In particular, we tested whether the diacylglycerol generated under these circumstances could induce the activation of protein kinase C. Serine phosphorylation of the CD3-γ chain is thought to occur in T cells as a consequence of the activation of PKC (7). Therefore, T cells were incubated with [³²P] orthophosphate, to label their endogenous ATP pool, and then stimulated with Ag and APC. Subsequently, plasma membranes were solubilized in detergent and the CD3 complex was immunoprecipitated. Finally, CD3-γ chain phosphorylation was examined following the fractionation of the immune complexes by nonreduced sodium dodecyl sulfate–polyacrylamide gel electrophoresis (SDS-Page) (7). Using this procedure, anergic T cells were examined for evidence of PKC activation following stimulation with antigen and APC. In several experiments, anergic T cells demonstrated poor antigen-induced incorporation of ³²P into CD3-γ pro-

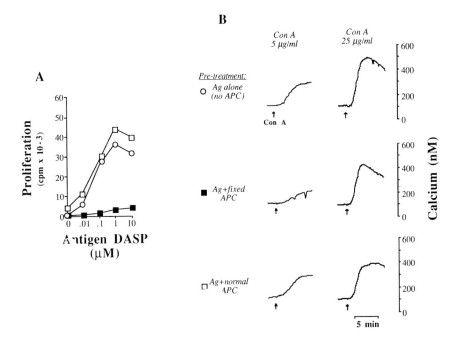

Fig. 6. Proliferation and increases in $[Ca^{2+}]_i$ of resting, preactivated, or anergized T cells. Resting A.E7 cells were pretreated overnight with 5 μM DASP with either no APC (*open circles*), 3000 rad–irradiated APC (*open squares*), or paraformaldehyde-fixed APC (*filled squares*). Following this pretreatment, viable T cells were separated from APC on Ficoll Hypaque gradients and placed back in culture with no additions. Six days later, cells from all three groups were harvested and dead cells and debris were removed on Ficoll Hypaque gradients. Proliferation in response to Ag and APC was measured, as described in the legend of Fig. 5, at doses of Ag indicated in (A). For $[Ca^{2+}]_i$ measurements, the T cells were loaded with Fura-2 AM (2 μM) for 30 min at 37°C and then washed. Cells were stimulated with either 5 or 25 μg/ml Con A, and $[Ca^{2+}]_i$ was determined by monitoring the increase in the ratio of fluorescence at 550 for excitation at 340 and 380 nm. Arrows indicate time at which stimulus was added.

tein. As illustrated in Fig. 7A, one experiment showed CD3-γ chain phosphorylation to be extremely weak in T cells following the induction of anergy when compared to normal control T cells. This could have suggested a defect in the activity of PKC associated with the anergic state; however, this interpretation was not compelling for several reasons. First, in this and several other experiments, the general rate of ^{32}P incorporation was low, as evidenced by the greatly reduced phosphorylation of the class I molecule immunoprecipitated from these anergic T cells. The reason for this general decrease in apparent protein phosphorylation remains uncertain; however, a reduction in the ATP

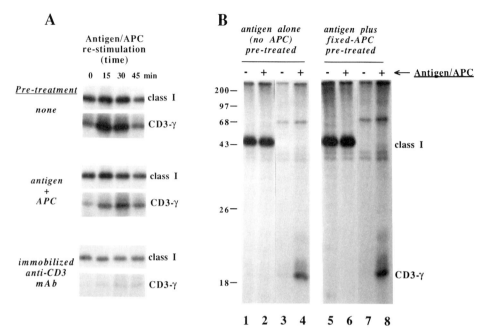

Fig. 7. CD3-γ chain phosphorylation in anergic T cells. (A) A.E7 T cells were incubated overnight with either medium alone, B10.A spleen cells plus pigeon cytochrome *c* (100 μg/ml), or immobilized anti-CD3 mAb 145-2C11 (1 μg/ml). Seven days later, viable T cells were recovered, and each group was incubated with 3 mCi of [^{32}P]orthophosphate in phosphate-free RPMI 1640 medium plus 10% dialyzed fetal calf serum for 3 hours. Following labeling of the ATP pool, washed T cells (12.5 × 10^6) were stimulated with syngeneic B10.A spleen cells (5 × 10^7) prepulsed with 100 μM fragment 81–104 for either 0, 15, 30, or 45 min. After the stimulation, CD3-γ chain phosphory-lation was determined. (B) A.E7 T cells were initially incubated 48 hours with fragment 81–104, ei-ther without APC or in the presence of paraformaldenhyde-fixed B10.A spleen cells, and then allowed to rest in the absence of stimuli for 2 weeks. T cells were recovered, loaded with [^{32}P]or-thophosphate (3.3 mCi each group for 2 hours), and then stimulated with B10.A spleen cells with or without fragment 81–104 at a concentration of 100 μM (6 × 10^6 T cells and 30 × 10^6 APC per group). After a 45-min incubation, phosphorylated CD3-γ and class I heavy chain were measured. *Methods:* Protein phosphorylation was measured as previously described (7). Briefly, cells were lysed in 0.5% Triton X-100 lysis buffer containing phosphatase and protease inhibitors at 4°C, and nuclei were removed by centrifugation. Lysates were serially incubated with protein A–Sepharose beads alone, beads coated with the anti-CD3 mAb 145-2C11, and then beads coated with the anti–class I mAb AF3. Immunoprecipitated class I and CD3 antigens were eluted from the beads by boiling 5 min in SDS sample buffer and analyzed unreduced by SDS-PAGE on a 16.5% gel fol-lowed by autoradiography. In (A), only the regions of the gel corresponding to the phosphorylated class I heavy chain (45 kDa, *top*) and CD3-γ chain (20 kDa, *bottom*) are shown for each of the groups. In (B), lanes 1, 2, 5, and 6 are the immunoprecipitates with anti–class I antibody and lanes 3, 4, 7, and 8 are the immunoprecipitates with anti-CD3 antibody.

pool size, or decrease in the specific activity of the ^{32}P label within the ATP pool, could account for such a change. Alternatively, the development of clonal anergy in these cells may affect ATP metabolism.

A second reason for discounting the observed reduction in the level of CD3-γ phosphorylation is that normally preactivated T cells (pretreated with Ag/APC) also tended to show reduced stimulation of the γ chain phosphorylation (Fig. 7A). Again, this result might be most consistent with a state of antigen–receptor desensitization following a recent stimulation. Finally, in some experiments, general levels of protein phosphorylation were observed to be nearly normal in the anergic T cell population, and in these experiments γ chain phosphorylation in the anergic T cell was intact (Fig. 7B). Taken together, these experiments failed to document a defect at the level of antigen-inducible PKC activity that correlated with the development of the anergic state and pointed toward an effect of anergy on other parallel signaling systems or a more distal signaling event.

As one test for an alternative signaling pathway, CD3-ζ chain tyrosine phosphorylation was examined in preliminary experiments with anergic cells, because the phosphorylation of this protein is thought to reflect the actions of tyrosine kinase and/or phosphatase activities in T cells. In the experiment shown in Fig. 8, CD3-ζ chain tyrosine phosphorylation is apparent based on the detection of a 20-kDa tyrosine-phosphorylated polypeptide in a Western blot following immunoprecipitation of the tyrosine-phosphorylated proteins from detergent lysates. This experiment demonstrates that tyrosine kinase or phosphatase activities can be up- (or down-) regulated in anergic T cells and again confirms the ability of the TCR/CD3 complex to function in anergic T cells. Rather than a proximal signal transduction defect, these data suggest a more distal defect in the anergic T cells.

REVERSAL OF ANERGY

It has been reported that stimulation of anergic cells with IL-2 causes loss of anergy (4, 15). We obtained similar results, that is, restoration of IL-2 production in cells anergized and then stimulated to divide with exogenous IL-2 (see Fig. 9B). We also observed that the degree of anergy decays with time in culture (compare open squares to open circles in Figs. 9A and B). In Fig. 9A, 7 days after anergy induction there is a 70-fold decrease in IL-2 production, whereas 13 days after anergy induction there is only a 10-fold decrease. One possible explanation for these results is preferential outgrowth or survival of nonanergized cells in the anergized population. Preliminary results on the fold expansion in response to recombinant IL-2 and the viability of anergic and rested controls showed no difference, arguing against this possibility. Further

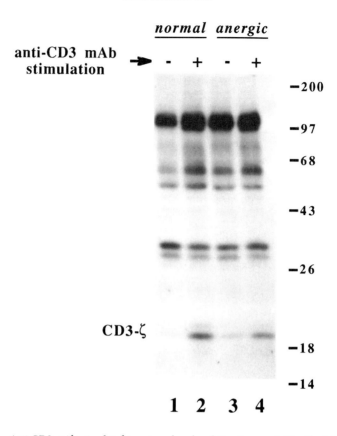

Fig. 8. Anti-CD3 mAb-stimulated tyrosine phosphorylation in anergic T cells. A.E7 T cells were initially incubated in medium alone or stimulated overnight with anti-CD3 mAb 145-2C11 (1 μg/ml immobilized) in the absence of accessory cells. After an 8-day rest period, viable T cells from each group were recovered, mixed 1:1 with T cell–depleted spleen cells, and then incubated (20 × 10⁶ T cells per group) for 30 min either with or without soluble anti-CD3 mAb 145-2C11 at a concentration of 1 μg/ml. *Methods:* To determine the level of anti-CD3 mAb-induced tyrosine phosphorylation of cellular proteins, cells were lysed in Triton X-100 lysis buffer containing phosphatase and protease inhibitors, and nuclei were removed. Lysates were then incubated with antiphosphotyrosine mAb PY20/protein A–Sepharose complexes. Immunoprecipitated phosphotyrosine-containing proteins were then eluted from the beads by boiling in sample buffer plus 5% 2-ME and subsequently separated by SDS-PAGE on a 10% gel. Proteins were transferred to a nitrocellulose membrane and probed with an antiphosphotyrosine rabbit antiserum. Antigen–antibody complexes were detected by incubation of the blot with ¹²⁵I-labeled protein A followed by autoradiography.

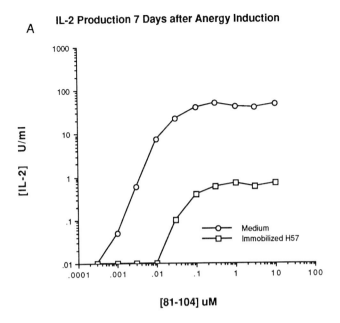

A IL-2 Production 7 Days after Anergy Induction

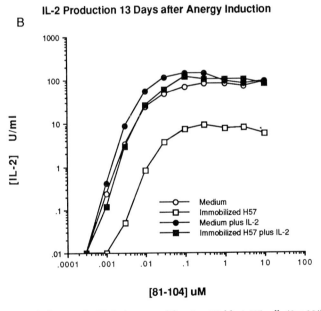

B IL-2 Production 13 Days after Anergy Induction

Fig. 9. Reversal of anergy by IL-2–driven proliferation. Viable A.E7 cells (2×10^4) pretreated (A) 8 days or (B) 14 days with (*open symbols*) or without (*filled symbols*) immobilized anti-TCR antibody (H57, 3 µg/ml) for 24 hours followed by stimulation with (*circles*) or without (*squares*) human recombinant IL-2 (30 U/ml) were stimulated with Ag and APC as described in the legend to Fig. 1. At 24 hours, 100 µl of supernatant was removed and duplicate samples were pooled and analyzed for IL-2 as described in the legend to Fig. 1. The limit of detection of the assay was 0.01 U/ml, and samples scoring negative are represented as 0.01 U/ml. No IL-2 was detected in any group in the absence of Ag.

experiments comparing precursor frequencies of IL-2 producers and responders and examination of IL-2 production by subclones obtained after IL-2 stimulation of both groups will be required to rule out definitively the outgrowth hypothesis. The slow loss of anergy on prolonged culture is also important in that it predicts that, if anergy maintains peripheral tolerance, it must be continually reinduced.

CONCLUSION

The major characteristic of anergic cells is a proliferative defect in response to stimulation by Ag and APC. This proliferative defect is due to a dramatic decrease in IL-2 production. In addition, anergic cells produce less of the other lymphokines, but this decreased production is graded and not as profound as that for IL-2. The decrease in lymphokine is correlated with decreased steady-state mRNA levels for these lymphokines.

The defect observed in anergic cells cannot be explained by a decrease in TCR expression or signaling mediated by the TCR as measured by inositol phosphate generation, increases in $[Ca^{2+}]_i$, or PKC activation. The effect of anergy on the activation of tyrosine kinases has not been thoroughly studied, but the preliminary results on tyrosine phosphorylation of the ζ chain of the TCR support the conclusion that anergy is not the result of impaired TCR signaling.

The finding that anergy is not permanent and can be reversed by IL-2–driven proliferation raises many interesting questions. Two that are extremely important with respect to tolerance in the periphery are (1) whether anergy must be constantly reinduced and (2) whether autoreactive T cells could be rescued by IL-2 from other T cells during an active immune response. Further studies on the nature of the induction, maintenance, and reversal of anergy are still required before we fully understand this state and its role in tolerance.

REFERENCES

1. Mueller, D. L., Jenkins, M. K., and Schwartz, R. H. (1989) *Annu. Rev. Immunol.* **7,** 445–80.
2. Schwartz, R. H., Mueller, D. L., Jenkins, M. K., and Quill, H. (1989) *Cold Spring Harbor Symp. Quan. Biology* **54,** 605–610.
3. Jenkins, M. K., Pardoll, D. M., Mizuguchi, J., Quill, H., and Schwartz, R. H. (1987) *Immunol. Rev.* **95,** 113–135.
4. Tomonari, K. (1985) *Cell. Immunol.* **96,** 147–62.
5. Quill, H., and Schwartz, R. H. (1987) *J. Immunol.* **138,** 3704–3712.
6. Jenkins, M. K., and Schwartz, R. H. (1987) *J. Exp. Med.* **165,** 302–319.
7. Mueller, D. L., Jenkins, M. K., and Schwartz, R. H. (1989) *J. Immunol.* **142,** 2617–2628.
8. Jenkins, M. K., Chen, C., Jung, G., Mueller, D. L., and Schwartz, R. H. (1990) *J. Immunol.* **144,** 16–22.

9. Gilbert, K. M., Hoang, K. D., and Weigle, W. O. (1990) *J. Immunol.* **144,** 2063–2071.
10. Stanley, E., Metcalf, D., Sobieszczuk, P., Gough, N., and Dunn, A. (1985) *EMBO J.* **4,** 2569.
11. Gray, P., and Goeddel, D. (1983) *Proc. Natl. Acad. Sci. USA* **80,** 5842–5846.
12. Ythier, A. A., Abbud-Filho, M., Williams, J. M., Loertscher, R., Schuster, M. W., Nowill, A., Hansen, J. A., Maltexos, D., and Strom, T. B. (1985) *Proc. Natl. Acad. Sci. USA* **82,** 7020–7024.
13. Torres, B. A., Farrar, W. L., and Johnson, H. M. (1982) *J. Immunol.* **128,** 2217–2219.
14. Weiss, A., Imboden, J., Hardy, K., Manger, B., Terhorst, C., and Stobo, J. (1986) *Annu. Rev. Immunol.* **4,** 593–619.
15. Essery, G., Feldmann, M., and Lamb, J. R. (1988) *Immunology* **64,** 413–417.

5

THE T CELL ANTIGEN RECEPTOR: BIOCHEMICAL ASPECTS OF SIGNAL TRANSDUCTION

LAWRENCE E. SAMELSON[1], JEFFREY N. SIEGEL[1],
ANDREW F. PHILLIPS[1], PILAR GARCIA-MORALES[1],
YASUHIRO MINAMI[1], RICHARD D. KLAUSNER[1],
MARY C. FLETCHER[2], AND CARL H. JUNE[2]

1. Cell Biology and Metabolism Branch
National Institute of Child Health and Human Development
National Institutes of Health
Bethesda, Maryland 20892
and
2. Immune Cell Biology Program
Naval Medical Research Institute
Bethesda, Maryland 20814

INTRODUCTION

In the early 1980s the dogma defining T cell function was that activation of protein kinase C (PKC) and elevation of intracellular calcium were sufficient for cellular activation (1). As was observed in many other cellular systems, investigators were able to induce T cell lymphokine production and T cell proliferation by the addition of phorbol esters and calcium ionophores. These investigators thus concluded that the T cell antigen receptor was coupled to phospholipase C (PLC) and that activation of this enzyme would induce polyphosphoinositide (PI) hydrolysis with subsequent generation of diacylglycerol and inositol phosphates. These second messengers would then lead to activation of protein kinase C and elevation of intracellular calcium. Receptor

55

regulation of these metabolites was viewed as critical to lymphokine gene transcription and cellular proliferation. Certainly, the activation of T lymphocytes by phorbol esters and calcium ionophores can be observed. However, there were some inconsistencies with this model. In the Jurkat system, investigators demonstrated that antigen receptor cross-linking by monoclonal antibody induced inositol phosphate production and calcium elevation. Presumably diacylglycerol was stoichiometrically produced after receptor cross-linking; nonetheless, investigators were forced to add phorbol esters to reconstitute full responses. Additional problems included the failure of the pharmacologic regimens to completely mimic the kinetics of T cell activation. There was always a concern that these pharmacologic agents might have multiple effects on the cells. The data that did most to question the adequacy of this model arose out of analysis of variants of the murine 2B4 hybridoma. Sussman *et al.* (2) were able to demonstrate a variant that produced interleukin-2 upon receptor cross-linking in the absence of demonstrable inositol phosphate release. Thus, although phosphoinositide breakdown is part of signaling in the T cell, its relationship to lymphokine production is not clear. Clearly, other pathways are activated through the T cell antigen receptor.

We took a different approach to questions of signal transduction in T cells. As of 1984 there had been many examples of ligand-induced receptor phosphorylation. For example, serine phosphorylation of the beta-adrenergic receptor induced by ligand had been demonstrated and related to receptor desensitization (3). Tyrosine phosphorylation of the epidermal growth factor (EGF) receptor and insulin receptors had been shown to be correlated with ligand binding and receptor kinase activation (4,5). In view of these studies pointing to the central role of protein kinase activitation in receptor-induced signal transduction, we asked whether the T cell antigen receptor was phosphorylated upon antigen engagement. To summarize several studies, we demonstrated that upon the addition of antigen, antireceptor antibody, or stimulatory anti–Thy-1 antibody, the antigen receptor is rapidly phosphorylated on two subunits (Fig. 1) (6–10). The CD3 γ chain is phosphorylated on serine residues, while the T cell receptor ζ chain is phosphorylated on tyrosine residues (11). The antigen-induced serine phosphorylation can be mimicked by addition of phorbol ester and is dependent on the presence of protein kinase C. The demonstration of tyrosine phosphorylation of a 21-kDa protein, later shown to be the T cell receptor ζ chain, was the first evidence for activation of a protein tyrosine kinase pathway in T lymphocytes. The receptor phosphorylations that we demonstrated could also be regulated by activation of a third kinase, the cyclic AMP–dependent protein kinase. Elevation of intracellular cyclic AMP levels resulted in inhibition of CD3 γ and T cell receptor ζ chain phosphorylation induced by antigen. This analysis of the phosphorylations occurring at the level of the receptor thus led to several insights. First, we demonstrated that

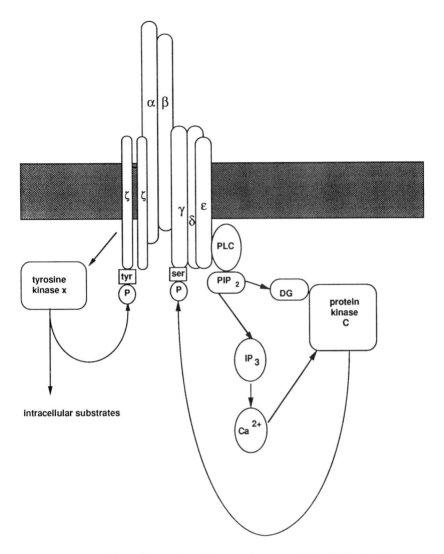

Fig. 1. A model defining the signal transduction pathways coupled to the T cell antigen receptor as described earlier (9).

multiple kinases were involved in regulation of receptor phosphorylation. Second, we confirmed that protein kinase C was activated through the antigen receptor, as had been predicted by other investigators. Third, we made the first demonstration of tyrosine kinase activation through the lymphocyte antigen receptor. It should be noted as well that since the antigen receptor subunits have now all been cloned and sequenced, it is clear that the antigen receptor itself cannot encode the tyrosine kinase. The T cell receptor system was thus the first example of receptor-mediated activation of a nonreceptor tyrosine kinase that must be associated with or coupled to the receptor. Finally, we addressed the question of whether the two kinase pathways, serine and tyrosine, are activated in parallel or in series. We demonstrated that depletion of protein kinase C had no effect on activation of the tyrosine kinase. However, this left open the question of whether the kinases are activated in parallel or whether the tyrosine kinase pathway could activate the serine kinase. This question has been addressed more recently (see below).

SUBSTRATE ANALYSIS IN MURINE AND HUMAN T CELLS; POSSIBLE TYROSINE KINASE REGULATION OF PHOSPHOLIPASE C

As described above, evidence for activation of a tyrosine kinase pathway in T cells was first provided by the observation that a 21-kDa protein associated with the antigen receptor is phosphorylated upon receptor occupancy. To further characterize the tyrosine kinase pathway in T cells, we have made extensive use of the technique of immunoblotting with specific antiphosphotyrosine antibodies. We have used this method to detect the phosphorylation of additional cellular substrates following receptor cross-linking (12). Comparison of lysates from unactivated and activated cells showed that phospho-ζ was detected in preparations from activated cells. In addition, we saw the phosphorylation of a 62-kDa protein and phosphorylation of proteins in the range 100–120 kDa. Further analysis by two-dimensional gel systems showed that the particular means of activation resulted in different patterns of substrate phosphorylation. Using the stimulatory anti–Thy-1 antibody, the ζ chain and 62-kDa substrates were the most prominent phosphoproteins. However, with antireceptor antibody or antigen, we detected relatively less 62-kDa phosphoproteins and more 53-kDa tyrosine phosphorylated proteins. The most impressive results of these studies were those of the kinetic analysis of tyrosine phosphorylation. Tyrosine phosphorylation of both the 62-kDa protein and the 53-kDa protein was very rapid with half-maximal phosphorylation by 2 min. Both half-maximal phosphorylation and the first detected phosphorylations preceded ζ chain tyrosine phosphorylation, which was not detected until 5 min and was half-maximal at about 15 min. This latter result suggested that the activation of tyrosine phosphorylation is very rapid and that, perhaps, phosphorylation of ζ chain represents a secondary and possibly regulatory event.

We have continued this characterization of the tyrosine kinase pathway by analyzing substrates in human T cells (13). We have used the Jurkat T cell tumor line and, in addition, highly purified populations of normal peripheral blood T cells. The advantage of the latter source is that results reflect the status of the pathway in highly homogeneous, quiescent, G_0 lymphocytes. In both the tumor line and the resting cells, the addition of anti-CD3 monoclonal antibodies results in a specific increase in tyrosine phosphorylation of a number of cellular proteins. A careful kinetic analysis demonstrated that phosphorylation of one of these proteins, a 135-kDa protein, could be detected as early as 5 sec after receptor ligation. The tyrosine phosphorylation of a 100-kDa substrate could be first detected at 15 sec and was already maximal at 45 sec. As observed in the murine system, tyrosine phosphorylation of the human T cell receptor ζ chain was much slower. The very rapid kinetics of substrate phosphorylation were compared to the kinetics of both intracellular calcium elevation and production of inositol phosphates. Both of these had been measured in many laboratories previously and were thought to be among the most rapid events following T cell receptor ligation. Direct comparison of calcium, PI turnover, and tyrosine phosphorylation of the substrates, however, revealed that phosphorylation of the 135- and 100-kDa proteins preceded detectable elevations in calcium or inositol phosphates. These kinetic relationships suggested that tyrosine phosphorylation might regulate inositol phosphate release. In this regard, the results were reminiscent of results discussed above for the EGF receptor and platelet-derived growth factor (PDGF) receptor systems, in which tyrosine phosphorylation of phospholipase C γ is thought to regulate enzymatic function (Fig. 2).

To address this question in T cells, we performed a series of studies with the tyrosine kinase inhibitor herbimycin (14). This drug was initially isolated as a reagent that could induce reversion of v-src–transformed cells. Addition of the reagent to T cells resulted in inhibition of the ability to induce tyrosine phosphorylation as detected by a lack of substrate phosphorylation. Surprisingly, the drug required at least 8 hours for maximal activity. The explanation for this time requirement was shown to be related to the effect of the drug on the level of tyrosine kinase protein. Treatment with herbimycin results in loss of lck protein, a process that might require this long incubation. The absence of kinase protein can obviously be detected as a loss of activity. Fyn and lck tyrosine kinase activities were reduced with the same kinetics and dose dependence as the inhibition of substrate tyrosine phosphorylation. The specificity of these effects was demonstrated by analyzing the level of c-raf serine kinase activity and protein. As will be described below, this kinase is activated in T cells by stimulation of protein kinase C. We demonstrated that there was a minimal effect on the level of PKC-induced raf activity and no loss of raf protein in response to overnight herbimycin treatment. This result demonstrates that there was minimal effect of the drug on the two serine kinases, PKC and raf. The functional effect of herbimycin treatment was then analyzed. Herbimycin was shown to inhibit

G PROTEIN   TYROSINE PHOSPHORYLATION

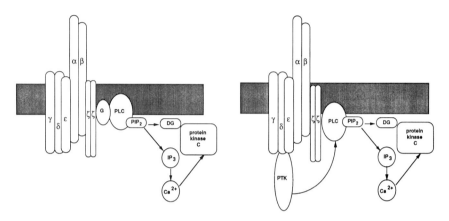

Fig. 2.   Alternative models for describing T cell antigen–receptor coupling to phospholipase C. The first activation of the receptor activates a guanine nucleotide-binding protein (G protein) that then interacts with phospholipase C, inducing its activation. In the second model we propose that activation of a tyrosine kinase results in tyrosine phosphorylation of phospholipase C or associated proteins, resulting in activation of the enzyme and polyphosphoinositide hydrolysis.

both inositol phosphate production and receptor-mediated calcium elevation. Addition of aluminum fluoride to these cells resulted in the expected induction of calcium elevation. Since this reagent presumably activates phospholipase C via a G protein, this result indicates that this enzyme is intact unless herbimycin directly inhibits the T cell receptor (TCR)–coupled PLC. We believe it more likely instead that tyrosine phosphorylation is required for activation of PLC via TCR ligation. As a further control, we analyzed the effect of herbimycin on such late T cell activation events as interleukin-2 (IL-2) production and IL-2 receptor expression. The former can be induced by the combination of phorbol myristate acetate (PMA) and ionomycin, while the latter, IL-2 receptor up-regulation, occurs with PMA alone. Neither of these transcriptionally regulated events was significantly affected by herbimycin. However, the drug entirely prevented CD3-induced IL-2 production and CD3-induced IL-2 receptor up-regulation. It is more difficult to ascribe inhibition of these late events to decreased tyrosine phosphorylation of substrates, but it certainly is a likely possibility. Thus, use of this drug further supports the kinetic arguments presented above; that is, tyrosine phosphorylation is important for phospholipase C activation or for phospholipase C regulation. Similar conclusions have been reached by Mustelin *et al.* (15) using the tyrosine kinase inhibitor genistein.

ANALYSIS OF TYROSINE PHOSPHATASES

Regulation of tyrosine phosphorylation by the action of tyrosine phosphatases has become of interest recently due to the recognition that CD45 is a tyrosine phosphatase (16). We have taken two approaches to the analysis of CD45 activity in T cells. In human T cells, as noted above, T cell stimulation induced via either the CD2 or CD3 molecules can be abrogated by coaggregation with the CD45 molecule using chemically cross-linked, heteroconjugate monoclonal antibodies (17). We have demonstrated that CD2 induces an increase in tyrosine phosphorylation in T cells and have confirmed that CD2–CD45 heteroconjugate antibodies inhibit calcium flux in these cells (18). We predicted that polyphosphoinositide hydrolysis would also be inhibited using these antibody heteroconjugates, and indeed we found that production of inositol phosphates IP_1, IP_2, IP_3 and IP_4 was markedly decreased. We then asked what the effect was on substrate tyrosine phosphorylation under these circumstances. CD2 stimulation results in tyrosine phosphorylation of 135-kDa and 100-kDa phosphoproteins, as was demonstrated with anti-CD3 stimulation. When one cross-links CD45 molecules via antibody-induced cross-linking, one sees an increase in tyrosine phosphorylation only of the 135-kDa protein. The same result is observed when one forms heteroconjugates between CD2 and CD45 with biotinylated antibodies and avidin. We detect an increase in phosphorylation of the 135-kDa protein without phosphorylation of the 100-kDa protein. We can perform the experiment in another fashion with the same result. In this case, one uses a pair of stimulatory anti-CD2 antibodies for activation and detects tyrosine phosphorylation of both substrates. When one adds to this combination anti-CD45 antibodies, which are biotinylated with avidin, one decreases the intensity and delays the onset of tyrosine phosphorylation of the 100-kDa protein. These results have several implications. First, we have provided further evidence that manipulation of the tyrosine phosphorylation pathway, this time by perturbation of a tyrosine phosphatase, affects phospholipase C activity. Second, we have demonstrated that tyrosine phosphorylation of a 100-kDa protein is correlated with PLC activity. Perhaps CD45 in some way regulates activity of the enzyme, for example, by controlling its level of tyrosine phosphorylation. Finally, the 100-kDa protein, perhaps, is a direct substrate for CD45.

A second approach to analyzing the role of tyrosine phosphatases was performed using the drug phenylarsine oxide (PAO). Investigators studying the insulin receptor system have shown that this reagent inhibits tyrosine phosphatases (19,20). When adipocytes or fibroblasts are preincubated with this drug prior to insulin treatment, one detects tyrosine-phosphorylated substrates that are otherwise not seen. The substrates presumably are tyrosine phosphorylated during ligand occupancy. However, normally, in the absence of phenylarsine oxide, tyrosine phosphate is turned over very rapidly. We asked what would happen when this drug was added to the murine hybridoma 2B4 (21). Initially, we demonstrated that PAO can inhibit tyrosine phosphatase activity of the CD45 molecule. CD45 was isolated with monoclonal antibodies, and the effect of phenylarsine oxide on its *in vitro* tyrosine phosphatase activity was tested. We demonstrated that PAO at concentrations between 5 and 10 μM resulted in 50% inhibition of phosphatase activity. We then demonstrated that the drug had no effect on the fyn or lck tyrosine kinases. Although there are certainly other tyrosine phosphatases and kinases in the cell, we use these data to support the assumption that the drug is working preferentially, if not solely, against tyrosine phosphatases and not kinases. The effects of the drug on the tyrosine phosphorylation of substrates are complex. Addition of PAO to unstimulated T cells results in a dose-dependent increase in tyrosine phosphorylation on a number of substrates that are otherwise not detected. This indicates either that there is a constitutively active tyrosine kinase in T cells or that PAO activates a tyrosine kinase. Of greater interest is the effect of the drug on stimulated T cells. At low concentrations, the drug appears to synergize with stimulation induced by the G7 anti-Thy-1 antibody. There is a dramatic increase in the induction of tyrosine phosphorylation of the T cell receptor ζ chain and of several other tyrosine kinase substrates of 54 and 62 kDa. The increase in phosphorylation of the 54-kDa substrate is nearly 14-fold. As the dose of PAO is increased, one observes inhibition of all activation-induced tyrosine phosphorylation. Thus, T cell receptor ζ chain tyrosine phosphorylation is completely inhibited, and the level of tyrosine phosphate on the 54- and 62-kDa substrates returns to the level induced by drug alone. These results have several implications. First, it appears that there is a constitutive level of tyrosine phosphatase activity in the cells which, when inhibited by low concentrations of PAO, results in detection of additional substrates and enhanced levels of tyrosine phosphorylation induced through the Thy-1 molecule. Second, the results of the experiments with the high concentrations of PAO indicate that tyrosine phosphatases may be involved in the regulation of tyrosine kinases. Inhibition of the phosphatase results in failure of tyrosine kinase activation, as assessed by substrate phosphorylation. In this way, the phenylarsine oxide experiments are compatible with the experiments of Pingel and Thomas, described above, in which the CD45 molecule is genetically absent (22). When CD45-negative cells are reconstituted with CD45 cDNA, TCR-mediated signaling is restored (23).

THE SERINE KINASE PATHWAY

The serine/threonine kinase c-*raf* has been identified as a substrate for the PDGF and EGF receptor tyrosine kinases. As described above, in a number of recent studies c-*raf* has been demonstrated to bind to activated PDGF or EGF receptor kinases upon ligand activation of those receptors. Moreover, c-*raf* has been shown to be phosphorylated on serine and, in some cases, tyrosine residues upon ligand addition. These phosphorylations induce activation of the c-*raf* kinase. We began to investigate c-*raf* as a possible substrate in the T cell system. We reasoned that since the T cell antigen receptor was coupled to both serine and tyrosine kinase pathways, c-*raf* could serve as a potential integrator of both kinases. We thus predicted that we could detect both serine and tyrosine phosphorylation of *raf*. We used specific anti-*raf* antibodies to immunoprecipitate *raf* from ^{32}P-labeled T cells that were either left unstimulated or stimulated with antireceptor antibodies, anti–Thy-1 antibodies, or phorbol esters (24). Under all these stimulatory conditions, we detected an increase in phosphate labeling of immunoprecipitated c-*raf* and a shift in c-*raf* migration on an SDS-PAGE gel so that it migrated with a greater apparent molecular weight.

To determine which kinase pathway activated c-*raf* phosphorylation, we performed phosphoamino acid analysis of isolated c-*raf*. We detected only serine and, to a lesser degree, threonine phosphorylation of c-*raf*. To confirm the absence of phosphotyrosine, we obtained the antiphosphotyrosine monoclonals used by Morrison *et al.* (25) to detect tyrosine phosphorylation of c-*raf* following PDGF treatment. These reagents were able to detect an increase in tyrosine phosphorylation on a number of substrates following T cell activation. One substrate was a 70-kDa protein which was a candidate for c-*raf*. However, anti-*raf* antibodies failed to immunoprecipitate any tyrosine-phosphorylated substrate. It thus seemed unlikely that the tyrosine kinase activated by the T cell antigen receptor was responsible for the hyperphosphorylation of *raf*. We then asked whether protein kinase C activation could be responsible. We depleted the cells of PKC with a high dose of phorbol ester and demonstrated that in cells treated in this fashion there was no further phorbol ester– or anti–Thy-1–stimulated *raf* phosphorylation. High-resolution phosphopeptide mapping was used to compare the pattern of phosphorylation induced by phorbol ester, anti–Thy-1, and antireceptor antibodies. A number of phosphopeptides were detected in the unactivated state, and several new ones were detected with stimulation. In all stimulatory circumstances, the pattern of phosphorylation was similar. In addition, in variants of 2B4 in which there was a failure of PI hydrolysis upon stimulation, there was a concomitant lack of *raf* activation.

The combination of phosphoamino acid analysis, phosphopeptide mapping, the analysis of 2B4 variants, and the PKC depletion experiments led us to the conclusion that the mechanism for phosphorylation of c-*raf* in T cells is due to activation of protein kinase C. The consequences of c-*raf* phosphorylation were then tested in *in vitro* kinase assays. Immunoprecipitates containing *raf* from unactivated and activated cells were incubated with histone H1 or H5 as exogenous substrates. c-*raf* isolated from activated cells induced an increase in phosphate incorporation into histone, as well as into a 120-kDa substrate which appears to coimmunoprecipitate with *raf*. Thus, stimulation of *raf* phosphorylation correlates with the increase in activity of the enzyme. In fact, dephosphorylation of the activated *raf* molecule by alkaline phosphatase resulted in inactivation of its kinase activity. These studies allow us to propose a model in which antigen receptor occupancy results in a cascade of events. Activation of phospholipase C presumably via the tyrosine phosphorylation pathway, as described above, leads to polyphosphoinositide hydrolysis, activation of protein kinase C with subsequent phosphorylation, and activation of the c-*raf* protein serine/threonine kinase. The function of activated *raf* remains to be determined. However, it is of note that activated c-*raf* appears to translocate to the nucleus, and activated forms of *raf* have been demonstrated to increase transcriptional activity of several genes.

FYN IS A T CELL RECEPTOR–ASSOCIATED TYROSINE KINASE

We have attempted to identify the kinase responsible for tyrosine phosphorylation of the T cell receptor ζ chain and phosphorylation of the other cellular substrates for several years. One candidate is lck, a member of the *src* kinase family. This tyrosine kinase is a T cell–specific kinase, which has recently been shown to be associated with the CD4 and CD8 molecules (26,27). However, for a number of reasons it is unlikely that lck is solely responsible for T cell receptor–induced tyrosine phosphorylation. For one, a number of cells are CD4 and CD8 negative. These include the γ-δ cells detected in thymus and periphery. In addition, the 2B4 hybridoma that we use extensively in our studies is currently CD4 negative. It is, of course, possible that lck is associated with yet another cell surface molecule such as the antigen receptor. Most problematic, however, is the observation that CD3 cross-linking, which results in T cell receptor activation and ζ chain phosphorylation, fails to result in lck activation (28). In contrast, cross-linking of CD4 or CD8 molecules results in lck activation, as demonstrated by increased autophosphorylation and phosphorylation of an exogenous substrate. Finally, the pattern of tyrosine-phosphorylated substrates differs whether one activates through the T cell antigen receptor or through the CD4 or CD8 molecule (29). For these reasons, we thought that yet another tyrosine kinase could be responsible for the tyrosine phosphorylations that we detect.

Several years ago we had attempted to demonstrate a kinase associated with the antigen receptor using immune complex kinase assays. These experiments were unsuccessful. However, about 1 year ago we repeated these experiments using digitonin instead of Triton X-100 as our means of solubilizing the antigen receptor. Immunoprecipitates using antireceptor antibodies were then incubated with kinase buffer containing radioactive [γ-^{32}P]ATP and the necessary cations for assaying protein tyrosine kinases. This protocol resulted in an impressive level of phosphorylation of several proteins specifically coimmunoprecipitated with antireceptor antibodies (30). The phosphoproteins have apparent molecular masses of 130, 120, 59, 56, 28, and 21 kDa. We failed to see a similar pattern of phosphoproteins when we used anti-H2, anti-LFA, or anti-CD45 antibodies. This pattern of phosphorylation was observed with anti–T cell receptor α antibodies (A2B4-2) or three anti-ε antibodies. Antibodies recognizing the carboxyl-terminal determinants of CD3-δ or T cell receptor ζ failed to immunoprecipitate proteins which were phosphorylated in a similar fashion. This result suggested that the kinase activity we were precipitating was associated with cytoplasmic domains of the antigen–receptor complex. We interpreted these results to indicate that a protein kinase was coprecipitated with the antigen receptor. We then determined that these phosphoproteins were predominantly phosphorylated on tyrosine residues. The lower-molecular-mass proteins (at 21 and 28 kDa) were exclusively tyrosine phosphorylated. The phosphoproteins at 56, 59, 120, and 130 kDa were phosphorylated on tyrosine and, to a lesser extent, serine residues. The identity of the lower-molecular-mass proteins was determined by two-dimensional diagonal gel analysis. The prominent phosphoproteins that we detected were phosphorylated T cell receptor ζ chains. The CD3 γ, δ, and ε chains were also tyrosine phosphorylated, but to a lesser extent.

To determine which protein tyrosine kinase was responsible for these tyrosine phosphorylation events, we compared the pattern of phosphoproteins that we detected with anti-CD3 ε antibodies and with a panel of antibodies binding to tyrosine kinases (27,31). By direct immunoprecipitation we determined that there were high levels of fyn and yes activity in these T cells and a low amount of lck activity. The pattern of phosphorylation that we detected with the anti-fyn antibodies was virtually identical to that which we saw with anti–CD3 ε precipitation. In particular, we observed phosphorylation of the 56-, 59-, 120-, and 130-kDa proteins, as well as low-molecular-mass proteins. When we looked at the phosphoproteins detected after anti-fyn immunoprecipitation on a two-dimensional diagonal gel, we again noted that we detected T cell receptor ζ chain phosphorylation as well as CD3 phosphorylation. Therefore, the pattern of phosphoproteins we detected with anti-TCR monoclonal antibodies was most like that seen with anti-fyn antibodies; moreover, anti-fyn antibodies immunoprecipitated the antigen receptor (Fig. 3). It should be noted that the fyn

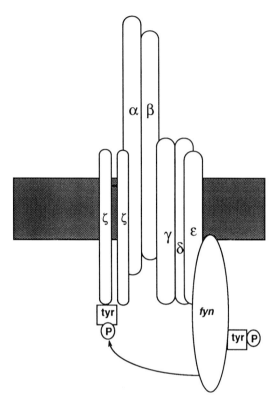

Fig. 3. The T cell antigen receptor is coupled to the fyn kinase. In *in vitro* kinase assays fyn is autophosphorylated and phosphorylates the antigen receptor ζ chain on tyrosine residues. There is no evidence to date that fyn is specifically associated with the CD3 ε chain. The depiction of the fyn–CD3 ε association in this model is purely arbitrary.

present in T cells differs from that in fibroblasts, suggesting that T cell fyn has a specific function (32).

Further confirmation that fyn is the kinase associated with the receptor came from two additional experiments. In one, we performed the *in vitro* kinase reaction after antireceptor antibody immunoprecipitation. This immune complex was then eluted with the detergent Nonidet P-40 (NP-40), which is known to disrupt the antigen receptor components. This detergent eluate was then subjected to reimmunoprecipitation with anti-lck and anti-fyn antibodies. We detected a 59-kDa phosphoprotein only with anti-fyn immunoprecipitation. Second, we performed a number of phosphopeptide analyses comparing the patterns of the phosphoproteins immunoprecipitated with antireceptor and anti-fyn antibodies. Digestion of the 130-, 120-, and 56-kDa phosphoproteins with staphylococcal V8 protease resulted in an identical pattern of phospho-peptides, whether one immunoprecipitated with anti-fyn or anti-receptor anti-bodies. The 59-kDa protein, which we believe is the fyn kinase itself, was also subjected to protease digestion. Fyns isolated directly from fibroblasts, from T cells, and via the antigen receptor from T cells share a number of peptides. There are subtle differences between the patterns of fyn isolated directly from the T cell and via the antigen receptor. We have performed further studies with trypsin digestion and high-resolution two-dimensional analysis. In these studies we demonstrate that both directly precipitated fyn and indirectly precipitated fyn share a common tyrosine-phosphorylated peptide, the site of autophospho-rylation. Fyn isolated from T cells via the antigen receptor contains several ad-ditional phosphopeptides which are phosphorylated on both tyrosine and serine residues. We are currently analyzing the stoichiometry of interaction and the question of whether fyn is activated via the antigen receptor.

REFERENCES

1. Weiss, A., Imboden, J., Hardy, K., Manger, B., Terhorst, C., and Stobo, J. (1986) *Annu. Rev. Immunol.* **4**, 593–619.
2. Sussman, J. J., Mercep, M., Saito, T., Germain, R. N., Bonvini, E., and Ashwell, J. D. (1988) *Nature* **334**, 625–628.
3. Sibley, D. R., Benovic, J. L., Caron, M. C., and Lefkowitz, R. J. (1987) *Cell* **48**, 913–922.
4. Cohen, S., Carpenter, G., and King, L., Jr. (1980) *J. Biol. Chem.* **255**, 4834–4842.
5. Kasuga, M., Karlsson, F. A., and Kahn, C. R. (1982) *Science* **215**, 185–187.
6. Samuelson, L. E., Harford, J. B., Schwartz, R. H., and Klausner, R. D. (1985) *Proc. Natl. Acad. Sci. USA* **82**, 1969–1973.
7. Samelson, L. E., Patel, M. D., Weissman, A. M., Harford, J. B., and Klausner, R. D. (1986) *Cell* **46**, 1083–1090.
8. Samelson, L. E., O'Shea, J. J., Luong, H. T., Ross, P., Urdahl, K. B., Klausner, R. D., and Bluestone, J. (1987) *J. Immunol.* **139**, 2708–2714.
9. Patel, M. D., Samelson, L. E., and Klausner, R. D. (1987) *J. Biol. Chem.* **262**, 5831–5838.
10. Klausner, R. D., O'Shea, J. J., Luong, H., Ross, P., Bluestone, J. A., and Samelson, L. E. (1987) *J. Biol. Chem.* **262**, 12654–12659.
11. Baniyash, M., Garcia-Morales, P., Luong, E., Samelson, L. E., and Klausner, R. D. (1988) *J. Biol. Chem.* **263**, 18225–18230.
12. Hsi, E. D., Siegel, J. N., Minami, Y., Luong, E. T., Klausner, R. D., and Samelson, L. E. (1989) *J. Biol. Chem.* **264**, 10836–10842.
13. June, C. H., Fletcher, M. C., Ledbetter, J. A., and Samelson, L. E. (1990) *J. Immunol.* **144**, 1591–1599.
14. June, C. H., Fletcher, M. C., Ledbetter, J. A., Schieven, G. L., Siegel, J. N., Phillips, A. F., and Samelson, L. E. (1990) *Proc. Natl. Acad. Sci. USA* **87**, 7722–7726.
15. Mustelin, T., Coggeshall, K. M., Isakov, N., and Altman, A. (1990) *Science* **247**, 1584–1587.
16. Charbonneau, H., Tonks, N. K., Walsh, K. A., and Fischer, E. H. (1988) *Proc. Natl. Acad. Sci. USA* **85**, 7182–7186.

17. Ledbetter, J. A., Tonks, N. K., Fischer, E. H., and Clark, E. A. (1988) *Proc. Natl. Acad. Sci. USA* **85**, 8628–8632.
18. Samelson, L. E., Fletcher, M. C., Ledbetter, J. A., and June, C. H. (1990) *J. Immunol.*, in press.
19. Bernier, N., Laird, D. M., and Lane, M. D. (1987) *Proc. Natl. Acad. Sci. USA* **84**, 1844–1848.
20. Levenson, R. M., and Blackshear, P. J. (1989) *J. Biol. Chem.* **264**, 19984–19993.
21. Garcia-Morales, P., Luong, E. T., Minami, Y., Klausner, R. D., and Samelson, L. E. (1990) *Proc. Natl. Acad. Sci. USA,* in press.
22. Pingel, J. T., and Thomas, M. L. (1989) *Cell* **58**, 1055–1065.
23. Koretzky, G. A., Picus, J., Thomas, M. L., and Weiss, A. (1990) *Nature* **346**, 66–68.
24. Siegel, J. N., Klausner, R. D., Rapp, U., and Samelson, L. E. (1990) *J. Biol. Chem.* **265**, 18472–18480.
25. Morrison, D. K., Kaplan, D. R., Escobedo, J. A., Rapp, U. R., Roberts, T. M., and Williams, L. T. (1989) *Cell* **58**, 649–657.
26. Rudd, C. E., Trevillyan, J. M., Dasgupta, J. D., Wong, L. L., and Schlossman, S. F. (1988) *Proc. Natl. Acad. Sci. USA* **85**, 5190–5194.
27. Veillette, A., Bookman, M. A., Horak, E. M., and Bolen, J. B. (1988) *Cell* **55**, 301–308.
28. Veillette, A., Bookman, M. A., Horak, E. M., Samelson, L. E., and Bolen, J. B. (1989) *Nature* **338**, 257–259.
29. Veillette, A., Bolen, J. B., and Bookman, M. B. (1989) *Mol. Cell. Biol.* **9**, 4441–4446.
30. Samelson, L. E., Phillips, A. F., Luong, E. T., and Klausner, R. D. (1990) *Proc. Natl. Acad. Sci. USA* **87**, 4358–4362.
31. Horak, I. D., Kawakami, T., Gregory, F., Robbins, K. C., and Bolen, J. B. (1989) *J. Virol.* **63**, 2343–2347.
32. Cooke, M. P., and Perlmutter, R. M. (1990) *New Biol.* **1**, 66–74.

6

Structure and Signaling Function of B Cell Antigen Receptors of Different Classes

MICHAEL RETH, JOACHIM HOMBACH, PETER WEISER,
AND JÜRGEN WIENANDS
Max-Planck-Institut für Immunbiologie
7800 Freiburg
Germany

INTRODUCTION

The antigen receptors of different immunoglobulin (Ig) classes (IgM, IgD, IgG, IgA, IgE) are expressed during B cell development in an ordered fashion. Early B cells express on their surface antigen receptors of the IgM class, and mature B cells coexpress IgM and IgD. The other Ig classes are expressed on the B cell surface only after a primary activation and a class switch event. After cross-linking by antigen or anti-Ig antibodies, the antigen receptor is activated and transduces signals to the cytoplasm. Depending on the developmental stage of the B cell, the signals can result in different biological responses. In early B cells they can result in cell death or apoptosis (1, 2), and in mature B cells they can result in either activation or deactivation (anergy) of the B cells (3–5).

How signals are generated from an activated antigen receptor is presently unknown, and the involvement of either G proteins or specific tyrosine kinases is discussed. The question of signal transduction has been especially puzzling because the membrane-bound Ig heavy chain of most antigen receptors lacks a cytoplasmic tail. We have discovered two new components (IgM-α and Ig-β) of the IgM antigen receptor (6–8). IgM-α and Ig-β are two transmembrane proteins forming a heterodimer which is specifically associated on the surface with

Molecular Mechanisms of
Immunological Self-Recognition

the membrane-bound IgM molecule. The α component of the heterodimer has a cytoplasmic tail of more than 60 amino acids. The antigen receptors of other Ig classes contain similar heterodimers, and our current knowledge of these receptors is summarized in this chapter.

RESULTS

A Heterodimer Is Found in the Antigen Receptor of Three Different Ig Classes

We have now analyzed the components of the antigen receptor of three different Ig classes (IgM, IgD, and IgG). Vectors expressing the membrane forms of the different Ig heavy chain classes were introduced together with a λ_1 light chain vector into the B lymphoma line K46. From these experiments we obtained K46 transfectants expressing either IgM, IgD, or IgG $_{2a}$ on their cell surface. The transfectants were either biosynthetically labeled or surface iodinated and then were lysed with 1% digitonin. The antigen receptor was isolated by affinity chromatography of the lysates and analyzed on two-dimensional sodium dodecyl sulfate (SDS) gels (7). All three receptors contained an α-β heterodimer. The β component (Ig-β) seems to be the same protein in all three heterodimers. It has a molecular mass of 39 kDa and can be most clearly identified after surface iodination (7). The α components seem to be isotype specific (8). They have molecular masses of 34 kDa (IgM-α), 35 kDa (IgD-α), and 33 kDa (IgG-α) and are most clearly seen in the analysis of biosynthetically labeled receptors.

The finding of different sizes for the α components of the three receptors suggested that IgM-α, IgD-α, and IgG-α are distinct proteins and that each of them is encoded by a separate gene. This notion was proved by the analysis of Ig-transfected J558L myeloma cells (Table I) expressing either μm (J558Lμm), δm (J558Lδm) or γ2am (J558Lγ2am) heavy chains. Myeloma cells as well as plasma cells secrete large amounts of antibody, and they normally express mRNA for the secreted form but not for the membrane form of Ig heavy chains. In the J558Lμm transfectant the mIgM molecules are not transported on the cell surface (9, 10). This is due to the fact that some of the genes for the α component of the antigen receptors are not expressed in myeloma cells (6). Amino-terminal protein sequencing (11) of IgM-α has shown that the gene encoding IgM-α is mb-1, a B cell–specific gene isolated by Sakaguchi et al. (12). The mb-1 gene and the IgM-α protein are expressed in pre-B and in mature B cells but not in myeloma cells. Once J558Lμm cells are cotransfected with an mb-1 vector (pSVmb-1) the IgM antigen receptor is expressed on the myeloma cell surface (7). In J558Lμm3 cells, an sIgM-positive variant of

TABLE I
Expression of α-β Heterodimers and sIg in Myeloma Cells

Cell line	Vector	α, β components	Surface Ig
J558L	—	β	—
J558Lμm	pSVμm	β	—
J558Lμm3	pSVμm	D-α, M-α, β	IgM
J558Lμm/mb-1	pSVmb-1	M-α, β	IgM
J558Lμm3/δm	pSVδm	D-α, M-α, β	IgM, IgD
J558Lδm	pSVδm	β	—
J558Lδm2.6	pSVδm	D-α, β	IgD
J558Lδm2.6μm	pSVμm	D-α, β	IgD
J558Lγ2am	pSVγ2am	G-α, β	IgG

J558Lμm which we isolated by cell sorting, mb-1 and IgM-α are expressed. Both experiments demonstrate the close correlation between mb-1, IgM-α, and sIgM expression. The B cell antigen receptor thus seems to be similar to the T cell antigen receptor (13, 14) in that it requires complete assembly from its different components IgM, IgM-α, and Ig-β prior to its surface expression.

The antigen receptor of the IgD class behaves similarly to that of the IgM class. In J558Lδm cells transfected with a vector for the membrane-bound form of the δ heavy chain (pSVδm) the IgD molecules are not transported on the cell surface (8). Using a cell sorter, it was possible to isolate the sIgD-expressing variant J558δm2.6. In this variant IgD molecules were expressed together with an α-β heterodimer on the cell surface. The α component of the IgD antigen receptor (IgD-α), however, was clearly different from IgM-α because the mb-1 gene is not expressed in J558Lδm2.6 cells. Consequently, in J558Lδm2.6 cells cotransfected with the pSVμm vector, IgD but not IgM molecules are expressed on the cell surface. Cotransfection of J558Lμm3 cells with the pSVδm vector, on the contrary, resulted in cells which coexpressed IgM and IgD. Thus, in the J558Lμm3 variant the genes for IgM-α and IgD-α were expressed, whereas in the J558Lδm2.6 variant only the gene for IgD-α was reactivated.

As a third major class of antigen receptor we analyzed the surface expression of IgG in myeloma cells (15). A vector for the membrane-bound form of IgG2a (pSVγ2am) was transfected into J558L cells. The obtained J558Lγ2am transfectants all expressed IgG$_{2a}$ molecules together with a α-β heterodimer on the cell surface. Thus, the gene for the IgG specific α component (IgG-α) seems to be constitutively expressed in J558L myeloma cells and as such is clearly different from the gene encoding IgM-α and IgD-α. In summary, it appears that each class of antigen receptor studied so far contains an α-β heterodimer with an

isotype-specific α subunit (Fig. 1). The finding of isotype-specific α components suggests a model in which the α components are mediating the class-specific contact between the α-β heterodimer and the Ig heavy chain. The mb-1
sequence (12) identifies IgM-α as a new member of the Ig superfamily. IgM-α
has an extracellular C domain which could be interacting specifically with the
last (membrane-proximal) C domain of the μ chain (Cμ4).

Experiments with various μm and δm vectors indicate that indeed the presence of only the membrane-proximal C domain and the corresponding transmembrane part are sufficient for the specific binding of the α-β heterodimer.
Exchange of the transmembrane part of μm with that of the H2K^K molecule results in loss of binding to the heterodimer (7). Thus the transmembrane parts
seem to be another contact area between IgM-α and μm. Because in the antigen receptor the Ig heavy chains form a homodimer with two identical surface
areas, the α component should bind to both sides of the Ig molecule. The receptor, therefore, presumably carries two α-β heterodimers (Fig. 1).

Signal Function of the Antigen Receptors of Different Classes

The K46 B lymphoma line expresses the genes of all three α components
studied. This is indicated by the finding that IgM, IgD, and IgG2a molecules
are expressed right away on the surface of the different K46 transfectants. Indeed, the genes for the different α components seem to be expressed early during B cell development. The WEHI 231 line represents an early B cell stage.

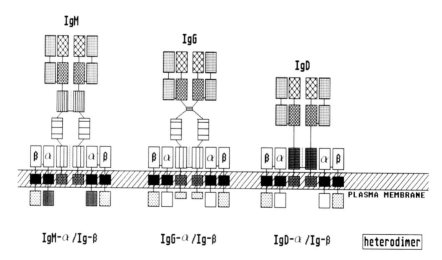

Fig. 1.   Structural model of the B cell antigen receptor of the IgM, IgG, and IgD classes.

This cell line expresses only IgM, whereas mature B cells coexpress IgM and IgD. After transfection of the pSVγ2am vector into WEHI 231 cells, the IgG$_{2a}$ antigen receptor is expressed on the cell surface (Fig. 2). The WEHI 231 line can also express the IgD antigen receptor (1, 16) indicating that it contains all three α components. The expression of the α component is thus independent of the Ig expression at the different stages of B cell development.

The α-β heterodimer may play a role in signal transduction by the activated antigen receptor. Indeed, the cytoplasmic part of IgM-α contains a conserved sequence motif (17) which is also found in the cytoplasmic part of the CD3 components of the T cell receptor and thus may play a role in signaling. The conserved sequence contains tyrosines which may be subjected to phosphorylation because the α and β components can be detected as phosphor protein in antigen receptors of splenic B cells (18). Although the primary events during antigen receptor signaling are still unclear, the signal transduction can be monitored via the production of the secondary messengers Ca^{2+} and inositol trisphosphate. With these assays similar responses are seen from all classes of antigen receptors (19). However, the biological response of B cells to antigen receptor cross-linking can be different depending on the class of antigen receptor activated.

The WEHI 231γ2am transfectant coexpresses IgM and IgG2a antigen receptors on the cell surface (Fig. 2). Cross-linking of the IgM antigen receptor with

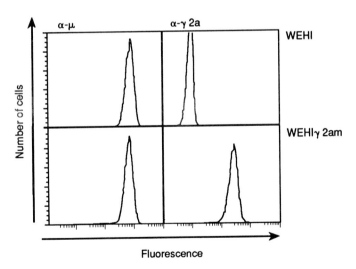

Fig. 2. Surface IgM and IgG2a expression of WEHI 231 and WEHI 231γ2am cells. Cells were stained with fluorescein-labeled goat antimouse IgM or IgG2a antisera (Southern Biotechnology, Birmingham, Alabama). Histograms were generated by the analysis of 10,000 cells in the FACscan.

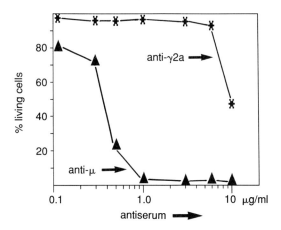

Fig. 3. Biological response of WEHI 231γ2am cells after cross-linking with goat anti-μ or anti-γ2a antibodies. Cells (10^6) were cultured for 1 day in increasing concentrations of antiserum and then counted to determine their viability.

anti-μ antibodies results in death of the WEHI 231 cells (1, 16). The cross-linking of the IgG_{2a} antigen receptor on the same cells, however, does not result in cell death (Fig. 3). The reduced viability of WEHI 231 cells in the presence of high concentrations (>5 μg/ml) of antibody is due to nonspecific cytotoxicity. The cross-linking of IgD antigen receptor of δ transfectants of WEHI 231 also does not result in cell death (1, 16). Thus different classes of antigen receptors can mediate different biological responses. The isotype-specific α component may thus play a role in the class-specific responses. By a structural and functional analysis of the α-β heterodimer in the different antigen receptors we hope to gain more insight into the transduction and regulation of the signals transmitted from the activated B cell receptor. This knowledge may lead to a better understanding of such important phenomena as B cell tolerance, B cell activation, and B cell memory.

REFERENCES

1. Tisch, R., Roifman, C. F., and Hozumi, N. (1988) *Proc. Natl. Acad. Sci. USA* **85,** 6914–6918.
2. Benhamou, L. E., Cazenave, P.-A., and Sarthou, P. (1990) *Eur. J. Immunol.* **20,** 1405–1407.
3. Cambier, J. C., and Ransom, J. T. (1987) *Annu. Rev. Immunol.* **5,** 175.
4. DeFranco, A. L. (1987) *Annu. Rev. Cell Biol.* **3,** 143.
5. Goodnow, C. C., Adelstein, S., and Basten, A. (1990) *Science* **248,** 1373–1379.
6. Hombach, J., Leclercq, L., Radbruch, A., Rajewsky, K., and Reth, M. (1988) *EMBO J.* **7,** 3451–3456.

7. Hombach, J., Tsubata, T., Leclercq, L., Stappert, H., and Reth, M. (1990) *Nature* **343,** 760–762.

8. Wienands, J., Hombach, J., Radbruch, A., Riesterer, C., and Reth, M. (1990) *EMBO J.* **9,** 449–455.

9. Sitia, R., Neuberger, M. S., and Milstein, C. (1987) *EMBO J.* **6,** 3969–3977.

10. Hombach, J., Sablitzky, F., Rajewsky, K., and Reth, M. (1988) *J. Exp. Med.* **167,** 652–657.

11. Hombach, J., Lottspeich, F., and Reth, M. (unpublished).

12. Sakaguchi, N., Kashiwamura, S., Kimoto, M., Thalmann, P., and Melchers, F. (1988) *EMBO J.* **7,** 3457–3464.

13. Clevers, H., Alarcon, B., Wileman, T., and Terhorst, C. (1986) *Annu. Rev. Immunol.* **6,** 629–662.

14. Ashwell, J. D., and Klausner, R. D., (1990) *Annu. Rev. Immunol.* **8,** 139–167.

15. Weiser, P. and Reth, M. (unpublished).

16. Ales-Martinez, J. E., Warner, G. L., and Scott, D. W. (1988) *Proc. Natl. Acad. Sci. USA* **85,** 6919.

17. Reth, M. (1989) *Nature* **338,** 383.

18. Campbell, K. S. and Cambier, J. C. (1990) *EMBO J.* **9,** 441–448.

19. Justement, L. B., Wienands, J., Hombach, J., Reth, M. and Cambier, J. C. (1990) *J. Immunol.* **144,** 3272–3280.

PART IV

T CELL TOLERANCE

7

Self–Nonself Discrimination by T Lymphocytes

HARALD VON BOEHMER[1] AND BENEDITA ROCHA[2]
1. Basel Institute for Immunology
CH-4005 Basel, Switzerland
and
2. Unité INSERM U-25 CNRS UA-122
Hôpital Necker
75015 Paris, France

INTRODUCTION

The question of self–nonself discrimination by the immune system is still awaiting a complete solution. The problem was recognized a long time ago and Owen's observation (1945) (1) in dizygotic cattle twins, which were chimeric with regard to their blood cells, indicated that self-tolerance was acquired rather than inherited. Ideas by Burnet and Fenner (1949) (2), which were extended by Lederberg (1959) (3), had a simple—possibly too simple—solution to the problem of self-tolerance: it was argued that clones of lymphocytes, each bearing a distinct antigen receptor, went through an early phase in development where contact with antigen was lethal rather than inducing effector function. Thus, tolerance to self was thought to be due to the elimination of self-reactive clones early during their development. This concept had two weaknesses in explaining all self-tolerance. First, it was difficult to imagine that all self-antigens could reach developing lymphocytes in primary lymphoid organs. Second, considering the enormous diversity of antigen receptors themselves, it was difficult to see how deletion of lymphocytes induced by diverse receptors on other lymphocytes would leave anything substantial behind fit to deal with foreign antigens (4). Possible solutions to these considerations were first, that clonal deletion was not the only mechanism of tolerance and second, that tolerance induction required a certain threshold amount of antigen which probably was not reached by most idiotypic antigen–receptor sequences. The main question was clearly not whether clonal deletion was the only way to

Molecular Mechanisms of
Immunological Self-Recognition

achieve tolerance, but whether clonal deletion existed at all as a central tolerance mechanism, possibly aided by some additional mechanisms responsible
for silencing already matured lymphocytes which had escaped clonal deletion
simply because they did not encounter their self-antigen early in development.

For decades this question was extremely difficult to address because of the
lack of clonotypic receptor markers. This was especially so for T lymphocytes,
where antigen binding by surface receptors could not be visualized. This left an
ambiguity in the interpretation of results obtained with most tolerance models.
Tolerance models by Billingham, Brent, and Medawar (1953) and Hasek
(1953) (5,6) confirmed that tolerance could be acquired. Neonatal tolerance
could not be established in all mouse strain combinations and it was not clear
whether the relative ease with which tolerance could be induced in some
neonatal mice reflected the immaturity of lymphocytes or simply their relatively small numbers. Conceptually more clear-cut were models in which hemopoietic stem cells developed in the presence of foreign antigen in
hemopoietic chimeras prepared by injecting stem cells into lethally X-irradiated histoincompatible hosts. In this model mature T lymphocytes were delib-
erately removed from the donor cells, and it turned out that this tolerance
model worked in all strain combinations. These experiments showed that there
was something special about developing lymphocytes with regard to tolerance,
as they could be adapted to tolerate a large variety of strong histocompatibility
antigens (7). The experiments could, however, not establish that clonal deletion
was the mechanism. The read-out in all the models depended on activation of
lymphocytes *in vivo* or *in vitro* and therefore did not establish whether tolerance was due to silence or absence of lymphocytes bearing specific receptors
for the tolerogen. More conclusive studies became feasible only in the mid-
1980s, when it was possible to raise antibodies specific for clonotypic receptors
on T lymphocytes (8–10). Even this technological advance was not sufficient to
yield conclusive results because the frequency of cells bearing one particular T
cell receptor (TCR) was extremely low such that one clone could not be visualized during lymphocyte development. This was different with cells specific for
so-called superantigens, which are recognized by particular V_β protein sequences shared by heterogeneous T cell receptors (TCRs). Here a relatively
large fraction of heterogenous T cells appeared specific for one superantigen
(11). Studies employing superantigens provided some clues to repertoire selection during T cell development (12) but the essential points became clear only
when the selection of a single TCR specific for "conventional" antigens, i.e.,
peptides bound by major histocompatibility complex (MHC) encoded molecules, was studied in T cell receptor transgenic mice (13). The latter studies
clearly established two rules of self–nonself discrimination by the immune system: they showed that tolerance could result from clonal deletion of immature
T lymphocytes being confronted with the specific peptide as well as the MHC

molecule early in T cell maturation (14,15). In addition, they showed that T cell maturation required recognition of the MHC molecule in the absence of the specific peptide early in T cell development (15–17). As these studies have been published, only a brief summary will be provided in the following in order to introduce the experimental system which was used to study further details of intrathymic selection as well as postthymic selection of the T cell repertoire.

NEGATIVE AND POSITIVE SELECTION OF A TRANSGENIC RECEPTOR SPECIFIC FOR THE MALE-SPECIFIC PEPTIDE PRESENTED BY CLASS I H-2D^b MHC MOLECULES

The α and β TCR genes from the CD4$^-$8$^+$ cytolytic B6.2.16 clone specific for the male-specific peptide presented by H-2D^b MHC molecules were cloned, and cosmids containing flanking sequences which harbored T cell–specific regulatory elements were injected into fertilized eggs (18–20). Transgenic mice which expressed both genes were crossed onto the SCID background in order to produce mice expressing essentially one TCR only (21). Negative selection (deletion) was studied by comparing male and female $\alpha\beta$ TCR transgenic SCID mice expressing H-2D^b MHC molecules. Positive selection was studied by comparing female $\alpha\beta$ TCR transgenic SCID mice expressing or lacking H-2D^b MHC molecules. The studies involved the analysis of thymocyte subpopulations in the various mice. In an oversimplified way, one can picture the development of thymocytes as going from CD4$^-$8$^-$ thymocytes, which begin to express TCRs, through receptor-positive but functionally incompetent CD4$^+$8$^+$ intermediates into CD4$^+$8$^-$ or CD4$^-$8$^+$ TCR-positive, functionally competent T cells (Fig. 1).

Figure 2 shows the analysis of the thymus of the experimental mice compared to a normal thymus by staining thymocytes with CD4 and CD8 antibodies. As is evident from Fig. 2, the thymus of male $\alpha\beta$ TCR transgenic SCID mice as compared to female littermates is largely devoid of CD4$^-$8$^-$ and CD4$^-$8$^+$ as well as CD4$^+$8$^+$ thymocytes but contains comparable numbers of CD4$^-$8$^-$ thymocytes. Because single positive and double positive thymocytes comprise ~90% of all thymocytes, the total number of thymocytes in male animals in only one-tenth of that in females. The interpretation of this result is that clonal deletion eliminates even the earliest CD4$^+$8$^+$ immature thymocytes which express the transgenic receptor.

When one compares female $\alpha\beta$ TCR transgenic SCID mice which either lack or express H-2D^b MHC molecules, one finds that thymuses from both types of animals contain CD4$^-$8$^-$ as well as CD4$^+$8$^+$ thymocytes but only those from the latter have CD4$^-$8$^+$ but not CD4$^+$8$^-$ mature thymocytes (Fig. 2). Here the interpretation is that the generation of mature thymocytes requires binding

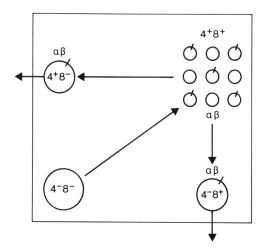

Fig. 1. A scheme of intrathymic development of cells expressing the αβ T cell receptor: Cycling CD4⁻8⁻ lymphoblasts differentiate into noncycling CD4⁺8⁺ and finally CD4⁺8⁻ and CD4⁻8⁺ thymocytes which express the αβ T cell receptor (symbolized by –).

of the αβ TCR to H-2D^b MHC molecules in the absence of the male-specific peptide. Furthermore, binding of the TCR to H-2D^b MHC molecules induces maturation of CD4⁻8⁺ thymocytes only, while binding to class II MHC molecules generates mature CD4⁺8⁻ thymocytes only (22,23). The latter conclusion is well supported by more recent independent experiments employing β1-microglobulin–deficient—and therefore class I MHC molecule–deficient—mice which have CD4⁻8⁻, CD4⁺8⁺, and CD4⁺8⁻ but lack CD4⁻8⁺ thymocytes (24).

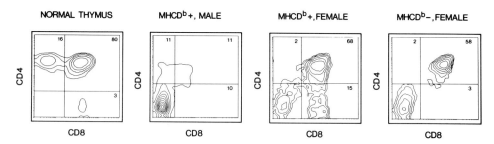

Fig. 2. The subset composition of the thymus from normal mice and four experimental mice by lymphocytes expressing CD4 and CD8 coreceptors. Numbers in quadrants represent the percentage of cells in these quadrants.

LACK OF ALLELIC EXCLUSION OF THE α TCR CHAIN IN αβ TCR TRANSGENIC MICE

The picture of thymocyte subsets in αβ TCR transgenic mice which were not crossed with SCID mice was slightly more complicated due to the fact that endogenous α but not β TCR genes could rearrange in these mice. In spite of the fact that the αβ TCR was expressed early on the surface of CD4⁻8⁻ thymocytes, endogenous α TCR genes were expressed at later stages at the RNA and protein level such that many cells contained transcripts from endogenous as well as the transgenic α TCR gene. Although in some cases this led to the expression of roughly equimolar amounts of endogenous and transgenic α TCR chains paired with the transgenic β TCR chain, most T cells expressed either high levels of transgenic or high levels of endogenous α TCR chain on the cell surface, presumably due to preferential pairing with the transgenic β chain. In the absence of productive rearrangements of endogenous α TCR gene segments, of course, only the transgenic α TCR chain was expressed. These αβ TCR transgenic mice contained significant numbers of CD4⁺8⁻ T cells in both the thymus (Fig. 3) and the periphery which all expressed high levels of endogenous α TCR chains at the cell surface. In contrast, most CD4⁻8⁺ thymocytes expressed high levels of the transgenic α TCR chain and a variable proportion of peripheral CD4⁻8⁺ T cells expressed high levels of endogenous α TCR chains. Thus, the early expression of the transgenic α TCR chain did not prevent endogenous α TCR rearrangement. In contrast, the early expression of the β transgene prevented rearrangement of V_β gene segments. Some of the receptors containing endogenous α TCR chains were selectable by thymic class I or class II MHC

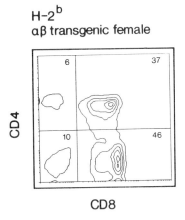

$H\text{-}2^b$
αβ transgenic female

CD4 / CD8 contour plot with quadrant values: 6, 37 (top), 10, 46 (bottom)

Fig. 3. The subset composition of the thymus from αβ T cell receptor transgenic mice not crossed onto the SCID background. Numbers in quadrants as in Fig. 2.

molecules and, therefore, mature T cells bearing these receptors could be detected in the thymus as well as peripheral lymph organs (13), and this has to be taken into consideration in the following experiments which were conducted with $\alpha\beta$ TCR transgenic mice which were not backcrossed onto the SCID background.

MUTATIONS IN THE PEPTIDE-BINDING GROOVE OF MHC MOLECULES AFFECT ANTIGENICITY AND NEGATIVE AS WELL AS POSITIVE SELECTION

Two naturally occurring mutants of the D^b MHC molecule have been analyzed in the bm13 and bm14 mutant strains. The latter contains a single mutation of the inward-pointing residue 70 (Glu → Asp) of the α_1 domain, while the bm13 contains three changed residues 114 (Leu → Glu), 116 (Phe → Tyr), and 119 (Glu → Asp) at the bottom of the groove (Fig. 4) (25). According to the Bjorkman–Wiley model (26), these residues appear to be inaccessible for the TCR and are considered to be involved in peptide binding. In contrast to cells from male B6 mice, cells from male bm13 and bm14 mice fail to stimulate the B6.2.16 clone from which the genes for the male-specific receptor were isolated. This agrees with earlier studies in these mice which showed that cells from female mutant mice could not be stimulated by cells from male littermates to produce male-specific cytolytic T lymphocytes (27). It was therefore expected that cells with the transgenic $\alpha\beta$ TCR in mice backcrossed onto the bm13 or bm14 background would show no deletion in animals homozygous for the bm13 or bm14 mutation. This was indeed the case, as shown in Fig. 5 for the bm13 homozygous male mice, which contain a high proportion of CD4+8+ thymocytes expressing intermediate levels of the transgenic receptor on the cell surface. In contrast, in male D^b/bm13 heterozygous mice this population was entirely deleted (25).

As positive selection proceeds in the absence of the specific peptide, one may or may not have anticipated that MHC mutations restricted to the peptide-binding groove would not affect positive selection. The fact that MHC molecules seldom exist on the surface of cells without a bound peptide (28) creates difficulties with the view that positive selection would be achieved simply by the binding of TCR to upward-pointing residues of MHC molecules without the positive or negative interference of the MHC-bound peptides. It may therefore not be surprising that indeed, as a rule, mutations which affect antigenicity also affect positive selection (25,29,30). This is shown in Figs. 6 and 7 for both the bm13 and the bm14 mutation. In both the periphery and the thymus of female D^b/bm13 heterozygous mice, one finds a discrete population of

bm 13

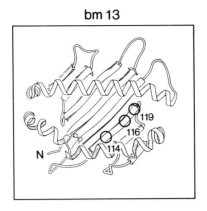

bm 14

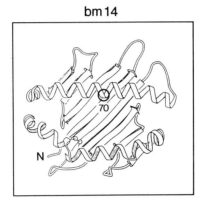

Fig. 4.   Position of residues in the mutant bm13 and bm14 MHC molecules which differ from the wild-type D^b MHC molecule.

CD4⁻8⁺ cells expressing a high level of the transgenic TCR visualized by the T3.70 antibody which is specific for the transgenic α TCR chain. The fact that this population is much smaller than in H-2D^b homozygous mice (Fig. 7) may indicate that various class I MHC molecules may compete for peptides or proteins which allow positive selection. Nevertheless, the selection of CD4⁻8⁺, T3.70 cells is clearly visible in D^b/bm13 heterozygous mice but absent in bm13 homozygous mice. Likewise, in hemopoietic chimeras produced by injecting T cell–depleted bone marrow cells from H-2^b αβ TCR transgenic female mice into lethally X-irradiated H-2D^b(B6) homozygous or bm14 homozygous recipient mice, it is evident that positive selection of CD4⁻8⁺, T3.70 cells occurs in the former but not the latter recipients. Unfortunately, these experiments do

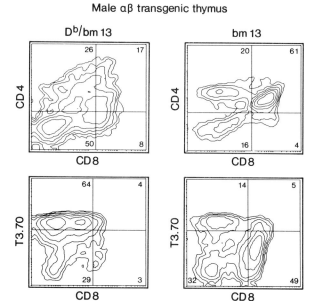

Fig. 5.   Absence of negative selection of male-specific thymocytes in the bm13 mutant strain. Note the absence of CD4⁺8⁺ T cells expressing low levels of the transgenic α TCR chain in the D^b/bm13 heterozygous but not bm13 homozygous mice.

not tell us whether the role of the peptide in positive selection is facilitating or interfering with positive selection by MHC molecules; they simply tell us that peptides are not ignored during positive selection.

POSITIVE SELECTION: EXPANSION OF MATURE THYMOCYTES OR MATURATION OF IMMATURE THYMOCYTES?

We have addressed the question of whether positive selection is reflected in expansion of mature T cells rather than maturation of immature resting cells by labeling experiments employing bromodeoxyuridine (BrdU), which is incorporated into newly synthesized DNA and can be visualized by a monoclonal antibody specific for BrdU (31). Continuous labeling experiments (BrdU injected intraperitoneally every day early in the morning and late in the evening) show that in female H-2^b αβ TCR transgenic mice, most CD4⁺8⁺ but none of the CD4⁻8⁺ thymocytes are labeled after 4 days of continuous labeling. This indicates that CD4⁻8⁺ thymocytes do not divide in these mice (Fig. 8). Whereas the

Female αβ transgenic thymus

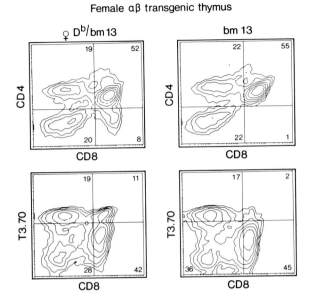

Fig. 6. Lack of positive selection in bm13 homozygous mice. Note the presence of the small population of CD8⁺, T3.70^high cells in Dᵇ/bm13 homozygous mice.

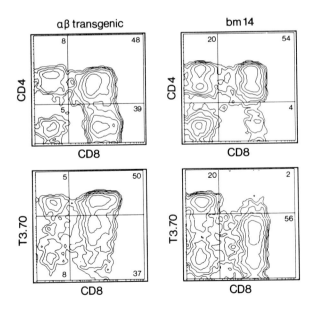

Fig. 7. Lack of positive selection in bm14 homozygous mice. Note the absence of CD8⁺, T3.70^high cells in the thymus of bm14 homozygous mice compared to αβ transgenic H-2ᵇ mice.

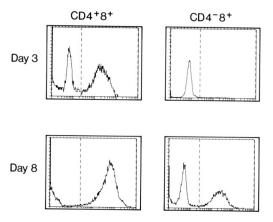

Fig. 8. The proportion of BrdU-labeled CD4⁺8⁺ and CD4⁻8⁺ cells after 3 and 8 days of continuous labeling.

label accumulates linearly in the CD4⁺8⁺ compartment during the first 5 days, it accumulates nonlinearly in the CD4⁻8⁺ compartment from day 5 onward (Fig. 9): at day 5 there is a steep increase in the proportion of labeled CD4⁻8⁺

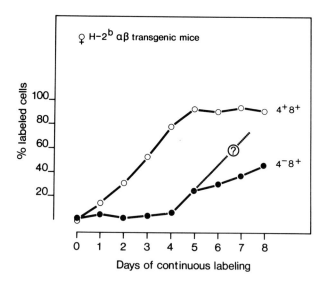

Fig. 9. The increase in percentage of labeled CD4⁺8⁺ and CD4⁻8⁺ during continuous labeling of female αβ TCR transgenic H-2ᵇ mice. The linear extrapolation of the line of labeled CD4⁻8⁺ cells may represent the real accumulation of labeled cells.

cells. Thereafter, the label accumulates more slowly. The nonlinear slope in the CD4⁻8⁺ compartment may be due to the fact that the time of migration from this compartment into the periphery is not related to the time of entry into it; i.e., the first cells which have entered are not necessarily the first ones to leave. This means that the accumulation of labeled cells in this compartment has kinetics that do not parallel the new generation of CD4⁻8⁺ thymocytes. If one extrapolates the line with the same slope as observed from day 4 to day 5, one can obtain two parallel lines indicative of new production of cells with the CD4⁺8⁺ and the CD4⁺8⁻ phenotype. Because approximately 40% of thymocytes are CD4⁻8⁺ cells in these mice, this could mean that almost all CD4⁺8⁺ precursors turn into CD4⁻8⁺ progeny. In any case, if one compares the absolute numbers of labeled CD4⁺8⁺ and CD4⁻8⁺ cells in αβ transgenic and normal B6 mice, it is clear that in the former the transition of CD4⁺8⁺ precursors into CD4⁻8⁺ progeny is much more efficient than in the latter (Fig. 10).

An independent experimental protocol leads to the same conclusion: if one reconstitutes X-irradiated H-2ᵇ mice with varying mixtures of T cell–depleted bone marrow cells from normal female B6 mice as well as H-2ᵇ female αβ TCR transgenic mice, one observes that the numbers of immature CD4⁺8⁺ cells expressing relatively low levels of the transgenic receptor and of CD4⁻8⁺ cells expressing high levels of the transgenic receptor are the same in each animal (Fig. 11, Table I). This is not true for cells expressing other receptors: here the total number of immature CD4⁺8⁺ is at least 10-fold higher than the number of

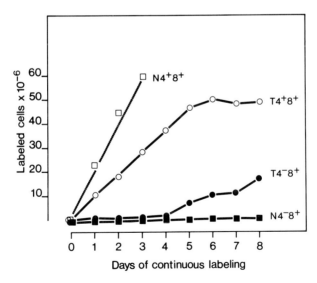

Fig. 10. The absolute numbers of labeled CD4⁺8⁺ and CD4⁻8⁺ thymocytes in normal (N) and αβ TCR transgenic (T) mice during continuous labeling with BrdU.

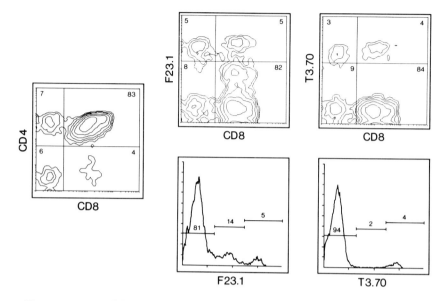

Fig. 11. Staining of thymocytes from hemopoietic chimeras prepared by injecting X-irradiated (800 R) female C57Bl/6 mice with a mixture of bone marrow cells from female C57Bl/6 mice (90%) and female αβ TCR transgenic C57Bl/6 mice (10%) by CD8, F23.1., and T3.70 antibodies. The F23.1 antibody detects the transgenic β TCR chain as well as $V_{\beta 5}$ TCR chain. The T3.70 antibody detects the transgenic α TCR chain only. The histogram shows the proportion of cells not staining at all, staining with intermediate intensity, or staining with high intensity with the F23.1 and T3.70 antibodies.

TABLE I

Proportion of Lymphocytes with Various Phenotypes in the Thymus of Different Bone Marrow Chimeras

	% of transgenic bone marrow cells		
Phenotype	100	90	50
CD4⁻8⁺	9	4	2
CD8⁺, F23.1^{low}	22	14	13
CD8⁺, F23.1^{high}	11	5	<1
CD8⁺, T3.70^{low}	10	2	0
CD8⁺, T3.70^{high}	9	4	0

CD4$^-$8$^+$ cells. Again, these experiments indicate that most of the immature CD4$^+$8$^+$ cells expressing the transgenic receptor mature into CD4$^-$8$^+$ cells. Collectively, these experiments leave very little doubt that the vast majority of immature CD4$^+$8$^+$ can be rescued from programmed cell death and induced to maturation provided these cells express a selectable receptor.

POSTTHYMIC EXPANSION OF MATURE T CELLS

Once mature T cells are formed, they can leave the thymus and undergo considerable expansion in peripheral lymph organs. The expansion potential of T cells has been shown in a variety of experiments, probably most directly by transferring small numbers of peripheral T cells into T cell–deficient, thymusless *nu/nu* mice, in which both CD4$^+$8$^-$ and CD4$^-$8$^+$ expand considerably (32). It is not clear whether this peripheral expansion randomly amplifies patterns of TCR specificities as they were selected in the thymus or whether it is not random and modifies patterns which were initially selected in the thymus. To study these questions, mature T cells from female αβ TCR transgenic mice were transferred into female and male *nu/nu* recipients. The remarkable observation in these experiments was that the entire increase in T cells in the T cell–deficient recipients could be attributed to CD4$^+$8$^-$ and CD4$^-$8$^+$ T cells expressing endogenous α TCR chains even though approximately half of the inoculated CD4$^-$8$^+$ T cells expressed the male-specific transgenic αβ TCR. In fact, these male-specific cells did not expand at all and the total number of cells with this phenotype was the same at day 5 and day 60 after transfer (Fig. 12a). This indicates that postthymic repertoire selection is totally different from thymic selection with regard to both specificity and mechanism: the former selects specificities of nontransgenic TCRs and involves expansion of cells; the latter selects predominantly the specificity of the transgenic TCR and involves maturation rather than expansion.

POSTTHYMIC TOLERANCE

We have begun to address the induction of unresponsiveness in mature T cells by transferring T cells from female αβ TCR transgenic B6 mice into male *nu/nu* B6 mice and by following the fate of male-specific T cells in these animals. As shown in Fig. 12b, these cells proliferate initially vigorously such that by day 5 after transfer the number of peripheral T cells is four times higher than that observed in female recipients injected with the same numbers of cells. After this, the number of male-specific CD4$^-$8$^+$ T cells declines steadily but significant numbers can still be detected even 8 weeks after transfer.

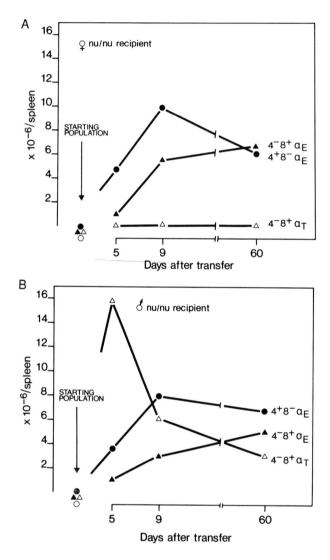

Fig. 12. (A) The total number of T cells in spleen and lymph nodes from female *nu/nu* mice injected with T cells from female αβ TCR transgenic H-2^b mice. α$_E$ means endogenous α TCR chains, α$_T$ transgenic α TCR chains. (B) The total number of T cells in spleen and lymph nodes from male *nu/nu* mice injected with T cells from female αβ TCR transgenic H-2^b mice.

When these cells are analyzed more carefully, one finds that their size is equal to that of cells with the transgenic receptor in female mice but that the levels of CD8 coreceptors as well as that of the transgenic receptor are lower in the maie than the female recipients (Fig. 13). In fact, the cells from the male recipients no longer proliferate in response to antigenic stimulation by male cells *in vitro*, even in the presence of interleukin-2 (IL-2), while proliferation of such male-specific cells is easily detected with cells from female recipients (Table II). Thus, as postthymic "positive selection" differs from thymic "positive selection," so does thymic "negative selection" differ from postthymic

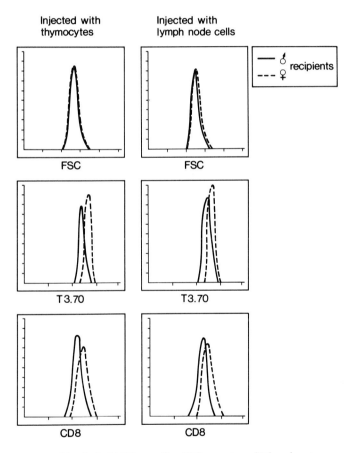

Fig. 13. Size and level of αβ TCR as well as CD8 receptors of T lymphocytes recovered from the lymph node of female (- - -) and male (–) *nu/nu* mice injected with T cells from female αβ TCR transgenic H-2[b] mice. Cells expressing CD8 and T3.70 molecules (double staining) were gated and single histograms were obtained from the gated cells.

TABLE II
Proliferative Response of T Cells from *nu/nu* Mice Injected with T Cells from Transgenic Mice[a]

Cells	+ IL-2			
	nu/nu ♂	*nu/nu* ♀	*nu/nu* ♂	*nu/nu* ♀
♀ αβ CD8	150,000	4,800	37,400	600
♀ αβ CD8 from *nu/nu* ♂ d5	74,500	35,200	17,500	5,000
♀ αβ CD8 from *nu/nu* ♂ d9	42,300	46,000	1,000	1,700
♀ αβ CD8 from *nu/nu* ♂ d20	7,400	5,800	600	700

[a] 5×10^4 responder cells (B cell– and CD4 cell–depleted lymph node cells) from various mice were cultured with 5×10^5 X-irradiated (3000R) spleen cells from male or female nude mice in the presence or absence of exogenous interleukin-2. Values represent the mean cpm of [³H]thymidine of triplicate cultures. Standard errors ≤10%.

"negative selection." Postthymic negative selection involves initial expansion and subsequent elimination of cells and/or down-regulation of antigen receptors as well as coreceptors; thymic negative selection involves deletion of cells at an immature stage without prior expansion.

DISCUSSION

Monoclonal antibodies specific for clonotypic TCRs and the construction of TCR transgenic mice have made it possible to study thymic (13–23) and postthymic T cell repertoire selection in previously unknown detail. The conclusions are that T cell maturation depends on "assertion" of self, namely binding of the αβ TCR to thymic MHC molecules in the absence of the specific peptide recognized by the receptor when expressed on mature T cells. This event is called positive selection, rescues CD4⁺8⁺ thymocytes from programmed cell death (33,34), and induces maturation into mature single positive thymocytes in the absence of cell division. An important feature of this selective step is that binding of the αβ TCR to class I or class II MHC molecules determines that the mature T cell is of the CD4⁻8⁺ and CD4⁺8⁻ phenotype, respectively (16,17,22,23).

In the presence of the specific peptide, one observes the opposite: CD4⁺8⁺ thymocytes are not even allowed to persist throughout their programmed life span but they are deleted as soon as they start to express the CD4/CD8 coreceptors (14,15). This event is called negative selection and prevents the entry of autoaggressive T cells into the peripheral lymphocyte pool.

Once the cells have passed through positive selection and avoided negative selection, they can expand considerably in peripheral lymph organs (32,35). This expansion depends again on the specificity of the αβ TCR but in a different way than when compared with intrathymic positive selection: here the receptor must bind to both the peptide and the presenting MHC molecule. The consequence of positive selection in peripheral lymph organs is also different from that in the thymus: in the former it is expansion and maturation into effector cells, while in the latter the consequence is differentiation into mature lymphocytes.

As predicted by Burnet and Fenner (1949) (2) and also by Lederberg (1959) (3), negative selection of mature T cells in peripheral or secondary lymph organs differs from that in primary lymph organs: in the thymus cells can be deleted without prior stimulation to expand, while in secondary lymph organs cells expand considerably but then either disappear and/or down-regulate antigen receptors and coreceptors such that the cells become refractory to antigenic stimulation even in the presence of exogenous IL-2. This mode of peripheral tolerance which we observe in male *nu/nu* mice injected with male-specific cells from αβ TCR transgenic mice differs from that seen in transgenic animals where certain antigens are expressed exclusively by "nonprofessional antigen"–presenting cells; in these experiments the antigens induced result a state of anergy in antigen-specific T cells which can be overcome by the addition of exogenous IL-2 (36). Thus, in addition to one central tolerance mechanism, there appear to be at least two and possibly more peripheral tolerance mechanisms, hopefully including some which are mediated by so-called suppressor cells.

Whereas the biological sense of negative selection of potentially autoaggressive cells is obvious, the biological meaning of positive selection has confused immunologists to some extent (37): the high frequency of alloreactive T cells, many of which probably recognize some peptides presented by foreign MHC antigens, in a repertoire selected by self-MHC antigens appears to make positive selection superfluous. It has been demonstrated in the past by experiments employing hemopoietic chimeras that this is not true: experiments in X-irradiated, MHC-homozygous animals reconstituted with cells from donors expressing either completely allogeneic or semiallogeneic MHC molecules have shown that positive selection has a considerable impact on "immunological fitness": immune responses in allogeneic chimeras [as long as one does not study responses to a multitude of allogeneic major or minor histocompatibility antigens (38)] are very much reduced even after *in vivo* priming (which, as we have shown in the transfer experiments reported here, can lead to enormous expansion of peripheral T cells); likewise, responses "restricted" by MHC molecules not encountered in the thymus are very much lower, or even absent, in semiallogeneic chimeras (39). Thus, after negative selection, responses restricted by MHC molecules not encountered in the thymus are as a rule orders of magnitude lower than those restricted by MHC molecules which were encountered in the thymus.

But does positive selection increase the frequency of self-MHC–restricted cells compared to the unselected repertoire? This question is difficult to address experimentally but the following consideration may indicate that this is so. From the data on positive selection of several transgenic αβ TCRs which have been studied so far, it is apparent that different MHC molecules rarely select the same αβ TCR and thus, in a first approximation, one can look at the unselected repertoire of immature T cells as a collection of 50 or so repertoires which can be selected by 50 different MHC molecules. Considering this assumption and the data on allogeneic or semiallogeneic chimeras, one comes to the conclusion that in the unselected repertoire, the frequency of self-MHC–restricted T cells will always be lower than in the selected repertoire. Thus, positive selection serves to increase the immunological fitness of the mature T cell pool in spite of some confusion raised by earlier experimental data. In summary, one may conclude that the experimental demonstration of allorestricted T cells has not been a good argument against the beneficial effect of positive selection but that the low frequency of T cells restricted by MHC antigens not present on the thymus in allogeneic or semiallogeneic chimeras has documented the usefulness of positive selection.

REFERENCES

1. Owen, R. D. (1945) *Science* **102**, 400.
2. Burnet, F. M., and Fenner, F. (1949) *The Production of Antibodies,* Macmillan, Melbourne, Australia.
3. Lederberg, J. (1959) *Science* **129**, 1649–1653.
4. Jerne, N. K. (1974) *Ann. Immunol.* **125c**, 373.
5. Billingham, R. E., Brent, L., and Medowar, P. B. (1953) *Nature* **172**, 603.
6. Hasek, M. (1953) *Ceckoslovenska. Biol.* **2**, 265.
7. von Boehmer, H., Sprent, J., and Nabholz, M. (1975) *J. Exp. Med.* **141**, 322.
8. Allison, J. P., McIntyre, B. W., Bloch, D. (1982) *J. Immunol.* **129**, 2293.
9. Meuer, S. C., Fitzgerald, K. A., Hussey, R. E., Hodgdon, J. C., Schlossman, S. F., and Reinherz, E. L. (1983) *J. Exp. Med.* **157**, 705.
10. Haskins, K., Kubo, R., White, J., Pigeon, M., Kappler, J. W., and Marrack, P. C. (1983) *J. Exp. Med.* **157**, 1149.
11. Marrack, P., and Kappler, J. (1990) *Science* **248**, 705.
12. Kappler, J. W., Roehm, N., and Marrack, P. C. (1987) *Cell* **49**, 273–280.
13. von Boehmer, H. (1990) *Annu. Rev. Immunol.* **8**, 531.
14. Kisielow, P., Teh, H. S., Blüthmann, H., and von Boehmer, H. (1988) *Nature* **335**, 730.

15. Sha, W. C., Nelson, C. A., Newberry, R. D., Kranz, D. M., Russell, J. H., and Loh, D. Y. (1988) *Nature* **335,** 271.

16. Teh, H. S., Kisielow, P., Scott, B., Kishi, H., Uematsu, Y., Blüthmann, H., and von Boehmer, H. (1988) *Nature* **335,** 229.

17. Kisielow, P., Blüthmann, H., Staerz, U. D., Steinmetz, M., and von Boehmer, H. (1988) *Nature* **333,** 742.

18. Uematsu, Y., Ryser, S., Dembic, Z., Borgulya, P., Krimpenfort, P., Berns, A., von Boehmer, H., and Steinmetz, M. (1988) *Cell* **52,** 831.

19. Krimpenfort, P., de Jong, R., Uematsu, Y., Dembic, Z., Ryser, S., von Boehmer, H., Steinmetz, M., Berns, A. (1988) *EMBO. J.* **7,** 745.

20. Blüthmann, H., Kisielow, P., Uematsu, Y., Malissen, M., Krimpenfort, P., Berns, A., von Boehmer, H., and Steinmetz, M. (1988) *Nature* **334,** 156.

21. Scott, B., Blüthmann, H., Teh, H. S., and von Boehmer, H. (1989) *Nature* **338,** 591–593.

22. Berg, L. J., Pullem, A. M., Fazekas de St Groth, B., Mathis, D., Benoist, C., and Davis, M. M. (1989) *Cell* **58,** 1035–1046.

23. Kaye, J., Hsu, M. L., Sauron, M. E., Jameson, J. C., Gascoigne, R. J., and Hedrick, S. M. (1989) *Nature* **341,** 746–749.

24. Zijlstra, M., Bix, M., Simister, N. E., Loring, J. M., Raulet, D. H., and Jaenisch, R. (1990) *Nature* **344,** 742.

25. Jacobs, H., von Boehmer, H., Melief, C. J. M., Berns, A. (1990). Mutations in the major histocompatibility complex class I antigen-presenting groove affect both negative and positive selection of T cells. *Eur. J. Immunol.* **20,** 2333–2337.

27. de Waal, L. P., Melvold, R. M., and Melof, C. J. M. (1983) *J. Exp. Med.* **158,** 1537.

28. Ljunggren, H., Stam, N. J., Oehlen, C., Neefjes, J. J., Höglund, P., Heemels, M., Bastin, J., Schumacher, T. N. M., Townsend, A., Kärre, K., and Ploegh, H. L. (1990) *Nature* **346,** 476.

29. Nikolic-Zugic, J., and Bevan, M. J. (1990) *Nature* **344,** 65–67.

30. Sha, W. C., Nelson, C. A., Newberry, R. D., Pullen, J. K., Pease, L. R., Russell, J. H., and Loh, D. Y. (1990). Positive selection of transgenic receptor-bearing thymocytes by K^b antigen is altered by K^b mutations that involve peptide binding. *Proc. Natl. Acad. Sci. USA* **87,** 6186–6190.

31. Dolbeere, F., Gretner, H., Pallovicini, M. G., and Gray, J. V. (1983) *Proc. Natl. Acad. Sci. USA* **80,** 5573.

32. Rocha, B., Dautigny, N., and Pereira, P. (1989) *Eur. J. Immunol.* **19,** 905–911.

33. von Boehmer, H. (1986) *Immunol. Today* **7,** 333.

34. von Boehmer, H. (1988) *Annu. Rev. Immunol.* **6,** 309–326.

35. Miller, R., and Stutman, O. (1984) *J. Immunol.* **133,** 2925–32.

36. Lo, D., Burkly, L., Flavell, R., Palmiter, R., and Brinster, R. J. (1989) *J. Exp. Med.* **169,** 779.

37. Matzinger, P. (1981) *Nature* **292,** 497.

38. Matzinger, P., and Mirkwood, G. (1978) *J. Exp. Med.* **184,** 84.

39. von Boehmer, H., Teh, H. S., Bennink, J. R., and Haas, W. (1985) In *Recognition and Regulation in Cell Mediated Immunity* (ed. Watson, J. D. and Mabrook, J.), Dekker New York, 89.

8

Transgenic Mouse Model of Lymphocyte Development

DENNIS Y. LOH
Departments of Medicine, Genetics,
and Molecular Microbiology
Howard Hughes Medical Institute
Washington University School of Medicine
St. Louis, Missouri 63110

INTRODUCTION

Thymus-derived lymphocytes, or T cells, play a critical role in an immune response. T cells can directly play the role of an effector cell or alternatively influence the function of the antibody-producing B lymphocytes. There are basically two kinds of T cells that have been well characterized. One kind is called cytotoxic T lymphocytes (CTL) or killer cells. Their main function is to recognize novel antigens on the surfaces of cells infected with viruses (or other microorganisms) and destroy them. The second kind is called helper T cells. Upon stimulation, they secrete specific cytokines, which in turn affect the differentiation of other T cells or more commonly B cells. In fact, when the antigen is protein in nature, the presence of the antigen-specific helper T cells is thought to be of critical importance in the development of the antibody-secreting plasma cells.

T cells recognize antigens as peptide fragments bound to products of self–major histocompatibility complex (MHC) molecules using the heterodimeric T cell receptor (TCR) (1, 2). This is in marked contrast to the B cell receptor, or the antibody molecule, which can recognize free antigens. Because T cells can recognize peptide antigens only in the context of self-MHC molecules which are present on the cell surface, T cell recognition occurs, by necessity, as cell–cell interactions. The phenomenon by which T cell recognition can occur only in the context of self-MHC molecules is called MHC restriction (3, 4). Most interestingly, self-MHC recognition is acquired during thymocyte

Molecular Mechanisms of
Immunological Self-Recognition

development as the TCR interacts with the self-MHC molecules. Another process that is of crucial importance during thymocyte development is that the T cells that emerge from the thymus must be rendered immunologically self-tolerant, since one does not desire those T cells that will be activated by self-antigens. If the self-reactive T cells are not eliminated, an autoimmune state may result. Thus, the selection criteria for the normal development of T cells are that they must be rendered self-MHC restricted and self-tolerant. It is the understanding of this process that this chapter will focus on.

T CELL RECEPTOR TRANSGENIC MOUSE AS A MODEL SYSTEM TO STUDY T LYMPHOCYTE DEVELOPMENT

Antigen recognition by the T cell is accomplished by the TCR. TCR molecules are composed of two chains, designated α and β, each of which has variable (V) and constant (C) regions (5). In the mouse, we had earlier shown that the repertoire of the V_β gene segments is relatively restricted (6). During development, the V gene segments for both chains undergo DNA gene rearrangement to complete the assembly of the heterodimeric TCR. Because these rearrangements appear to take place at random, stringent selection must take place subsequently to ensure the functional characteristics of the T cell population.

In normal mice, the TCR diversity is so large that it is virtually impossible to track the fate of individual cells (7). To overcome this limitation, we have decided to take a transgenic mouse approach. The strategy is to molecularly clone a pair of TCR genes that recognize a specific antigen and microinject it into one-cell mouse embryos. If the transgene is expressed at any significant level, one should be able to track its fate by use of a monoclonal antibody directed uniquely against the transgenic TCR and thus allow studies on T cell development. We have chosen an alloreactive cytotoxic T lymphocyte called 2C which originated in an H-2^b mouse and has specificity directed against the L^d component in the H-2^d background (8). This clone was chosen because there were monoclonal antibodies that could uniquely recognize the 2C β chain alone and 2C $\alpha\beta$ chain jointly. The availability of these reagents allowed us to track unequivocally the fate of the cells expressing the transgenic TCR.

After many failed attempts to create a transgenic TCR mouse that expressed the exogenous gene at any significant level, we finally succeeded by utilizing large cosmid constructs for both the α and β chains. Initial characterizations revealed that the transgene was expressed in an appropriate tissue-specific manner (9).

CLONAL DELETION OF SELF-REACTIVE T CELLS

Once the 2C TCR mice were made in the H-2^b background, they were back-crossed to the H-2^d mice to test what would happen when the transgene TCR is self-reactive against elements in H-2^d (10). Whereas in the H-2^b mice, the transgenic TCR$^+$ cells emerge into the periphery as CD4$^-$CD8$^+$ cells, no such cells could be detected in the H-2$^{b×d}$ mice. When the thymus was examined, there were no transgene-bearing cells in the CD4$^+$CD8$^+$ or in the single positive populations. These results strongly suggest that clonal deletion is taking place as a result of elimination of the self-reactive transgene-bearing thymocytes. The absence of the immature double positive (TCR^{lo}CD4$^+$CD8$^+$) cells strongly implies that it is at this stage that clonal deletion or negative selection is taking place. It is of fundamental importance that the immature cells appear to undergo clonal deletion as a result of interaction with the H-2L^d molecules, whereas the mature cells, as seen in the 2C clone, proliferate and are activated. This fundamental difference is of paramount interest in understanding the nature of thymocyte maturation.

POSITIVE SELECTION MODEL OF THE ORIGIN OF MHC-RESTRICTED T CELLS

With direct evidence obtained for clonal deletion as a possible mechanism for immunological self-tolerance, the 2C transgenic mouse was backcrossed to a third MHC haplotype. In the H-2^s background, the transgenic TCR-bearing cells failed to emerge into the periphery (10). Upon examining the thymus in the H-2^s mice, thymocytes appeared arrested at the TCR^{lo}CD4$^+$CD8$^+$ stage of development. This was in stark contrast to mice in the H-2^b background, where there was preferential emergence as CD4$^-$CD8$^+$ cells in the periphery. These data strongly suggested that the emergence of the 2C TCR-bearing cells required the presence of an element from the H-2^b background. The principle by which immature T cells require a specific interaction with self-MHC molecules is called *positive selection*. In our model, positive selection explains how all of our peripheral T cells are self-MHC restricted. For any of the T cells to emerge into the periphery, it must be able to interact with self-MHC molecules that are present in the thymus. This, by necessity, will skew the emerging T cell population toward self-MHC recognition. Thus the implication of the model is that thymocytes are "educated" by self-MHC molecules in the thymus, leading to preferential cell survival of these cells. By having stringent requirements for negative and positive selection, the peripheral T cell repertoire can satisfy the dual criteria of immunological self-tolerance and MHC restriction.

MOLECULAR REQUIREMENTS FOR POSITIVE SELECTION

The fundamental assumption underlying the positive selection model is that there is a requirement for a specific interaction between the TCR and the self-MHC molecules displayed in the thymus. If this were true, removal of the positive-selecting element should lead to abrogation of positive selection for given TCR-bearing cells. By using intra–H-2 recombinant mice, we mapped the positive-selecting element for the 2C TCR as H-2K^b. This was an unexpected result, since the alloreactive element that the 2C TCR recognizes is the H-2L^d molecule. It suggests that the origin of alloreactive cells can be any self-MHC molecule, not necessarily the allelic counterpart of the recognized allomolecule. It also strongly suggests that the origin of the alloreactive T cells is nothing but self-MHC positively selected cells that happen to "cross-react" with allo-MHC. Having mapped the gene locus (H-2K^b) whose product is responsible for the positive selection of the 2C TCR-bearing cells, we proceeded to formal testing by the equivalent of site-directed mutagenesis at this locus. Fortunately, there exists a series of mice congenic to the wild-type H-2^b mice in which the only difference is at the H-2K^b locus (11). These mice, called the bm mutant mice, carry specific DNA sequence changes at the K^b locus. Thus, by backcrossing the 2C TCR transgenic mice to the bm mutant mice, we would be formally testing the absolute dependence of the positive selection of 2C TCR by the H-2K^b molecule. When the TCR transgenic mice were put into the bm1 and bm10 backgrounds (these two mutants have sequence changes on the α2 helix of the K^b molecule), positive selection was totally abolished. This strongly suggests that there is a direct interaction between the 2C TCR and the K^b molecule that leads to positive selection. In other words, the interaction of these two molecules must transduce appropriate signals in the developing thymocytes that eventually lead to the differentiation that accompanies positive selection. Such processes include, but are not limited to, down-regulation of the CD4 gene as well as down-regulation of the recombinase (RAG1 and RAG2) and terminal transferase genes. In addition, the mechanism responsible for clonal deletion or programmed death of immature T cells must be turned off. Obviously, the elucidation of the signaling molecules responsible for this critical step in T cell development is of paramount importance.

In contrast to the bm1 and bm10 mutants, distinctively different results were obtained when the 2C TCR mice were backcrossed to the bm3 and bm11 mutants (both of which have mutations on the α1 helix, of which one position, 77, is shared). When the appropriate crossings were made, to our surprise, instead of the abolition of positive selection, negative selection or clonal deletion was observed. Moreover, the clonal deletion appeared to be restricted to the 2C TCR-bearing cells that had a relatively high level of the CD8 molecule. The CD8^{lo} cells appeared to emerge into the periphery as a result of positive selection. These data strongly suggest that the difference between positive and negative selection could be affected by changes in as little as one amino acid and that the level of the accessory molecule, CD8, can be a decisive factor in determining T cell fate (11).

FUTURE ISSUES IN T CELL DEVELOPMENT USING TCR TRANSGENIC MICE

The phenomena of positive and negative selection to account for the emergence of MHC-restricted, self-tolerant T cells have been well established. What remain to be solved are the exact cellular and molecular requirements responsible for these processes. Although there are many important facets that can be explored, I will focus on several specific problems that are of personal interest. One of the most fascinating aspects of T cell development is that the cell fate of a developing thymocyte appears to be determined by the nature of the interaction between the TCR and the self-MHC molecules in the thymus. The alternative pathways of cell lineage determination appear to be threefold. The first possibility is that the TCR interacts with the appropriate "strength" so that positive selection ensues. In this case, single positive (CD4 or CD8) cells emerge out of the thymus as functional, MHC-restricted cells. The second possibility is that the TCR interacts with the self-MHC molecules "too strongly," leading to the clonal deletion of such self-reactive cells. The third possibility is that the TCR fails to interact with the self-MHC molecules at all. In such cases, as seen in the 2C mouse in the H-2^s background, the developing thymocytes undergo maturational arrest at the double positive stage (TCR^{lo}CD4$^+$ CD8$^+$) in the thymus. It is presumed that these cells eventually die either in the thymus or shortly after exiting the thymus, since no double positive cells accumulate in the periphery. Thus, the fate of a given T cell appears to be entirely determined by the nature of the interaction between the TCR and the MHC molecules, since we assume that all of the other cell surface molecules are nonvariant. How does a cell know that its interaction with the self-MHC molecules is too strong such that it ought to be deleted? What is the signaling mechanism responsible for negative and positive selection? These are some of the most interesting questions that can be addressed using the transgenic TCR mice as a model system.

REFERENCES

1. Bjorkman, P. J., Saper, M. A., Samraoui, B., Bennett, W. S., Strominger, J. L., and Wiley, D. C. (1987) *Nature* **329,** 506–512.
2. Bjorkman, P. J., Saper, M. A., Samraoui, B., Bennett, W. S., Strominger, J. L., and Wiley, D. C. (1987) *Nature* **329,** 512–518.

3. Bevan, M. (1977) *Nature* **269,** 417–419.

4. Zinkernagel, R. M., and Doherty, P. C. (1974) *Nature* **248,** 701–702.

5. Gascoigne, N. R. J., Chien, Y. H., Becker, D. M., Kavaler, J., and Davis, M. M. (1984) *Nature* **310,** 387––391.

6. Behlke, M. A., Chou, H. S., Huppi, K., and Loh, D. Y. (1986) *Proc. Natl. Acad. Sci. USA* **83,** 767–771.

8. Kranz, D. M., Sherman, D., Sitkovsky, M., Pasternack, M., and Eisen, H. (1984) *Proc. Natl. Acad. Sci. USA* **81,** 573–577.

9. Sha, W. C., Nelson, C. A., Newberry, R. D., Kranz, D. M., Russell, J. H., and Loh, D. Y. (1988) *Nature* **335,** 271–274.

10. Sha, W. C., Nelson, C. A., Newberry, R. D., Kranz, D. M., Russell, J. H., and Loh, D. Y. (1988) *Nature* **336,** 73–79.

11. Sha, W. C., Nelson, C. A., Newberry, R. D., Pullen, J. K., Pease, L. R., Russell, J. H., and Loh, D. Y. (1990) *Proc. Natl. Acad. Sci. USA* **87,** 6186–6190.

9

Recognition Requirements for the Positive Selection of the T Cell Repertoire: A Role of Self-Peptides and Major Histocompatibility Complex Pockets

JANKO NIKOLIĆ-ŽUGIĆ[1,*] AND MICHAEL J. BEVAN[2]

1. *Immunology Program*
Sloan-Kettering Institute
New York, New York 10021
and
2. *Howard Hughes Medical Institute*
University of Washington
Seattle, Washington 98195

INTRODUCTION

T cells recognize fragments of foreign antigen in association with self major histocompatibility complex (MHC)-encoded molecules. In the thymus, the same MHC molecules shape the T cell repertoire by positive and negative selection (1–4). The interaction between the $\alpha\beta$ T cell receptor (TCR) of immature thymocytes and the MHC molecule expressed by thymic cortical epithelium is necessary for subsequent maturation of the thymocyte. Only cells

Please address all correspondence to: J. Nikolić-Žugić, Immunology Program, Memorial Sloan Kettering Cancer Center, 1275 York Avenue, New York, NY 10021.

Molecular Mechanisms of
Immunological Self-Recognition

expressing receptors which interact with sufficient affinity with the MHC-encoded molecules expressed on thymus epithelial cells are positively selected and go on to mature and seed the peripheral lymphoid organs.

The structure of a class I MHC molecule reveals a likely peptide binding groove composed of a "floor" of antiparallel β strands and "walls" of two long α-helices (5). It has been proposed that during MHC-restricted recognition of foreign peptide the T cell receptor makes contact with the outer face of the α-helices and the peptide lying between them (6–8). Since the selection of the T cell repertoire occurs during T cell differentiation in the thymus in the *absence* of foreign peptide, we speculated that the outer face of the α-helices would be particularly important in positive selection and the peptide binding groove would exert less (if any) influence. To address this we compared the ability of a number of H-2K^b variant MHC molecules (Table I) to select a K^b restricted repertoire. Most interest focused on two K^b mutants, H-2K^{bm8} (changes at positions 22, 23, 24, 30) and H-2K^{bm5} (position 116) with substitutions only in the β-strands forming the floor of the groove (9, 10). These changes are probably not detectable on the exterior of the molecule, and in fact these variants are serologically indistinguishable from wild-type K^b with a battery of antibodies (9, 11). Their effects on T cell recognition are likely to be due to changes in peptide binding (10). Comparison of the structure of HLA-A2 with that of a second class I molecule, HLA-Aw68, with 10 residue differences in positions lining the floor and walls of the groove shows that each substitution causes only very local structural changes (12). The backbone structure of the two molecules is the same and unaltered residues remain essentially unperturbed. For these reasons, the α-helical regions of K^{bm5} and K^{bm8} are expected to be identical to those on K^b. Two other K^b mutants, H-2K^{bm1} (positions 152,155,156) and

TABLE I
Characteristics of K^b Mutations

Mutant	Position of altered amino acid	Classification[a] of residue
K^{bm1}	152 Glu → Ala	Ligand/TCR
	155 Arg → Tyr	TCR/ligand
	156 Leu → Tyr	Ligand
K^{bm3}	77 Asp → Ser	Ligand
	89 Lys → Ala	Silent
K^{bm5}	116 Tyr → Phe	Ligand
K^{bm8}	22 Tyr → Phe	Ligand
	23 Met→ Ile	Silent
	24 Glu → Ser	Ligand
	30 Tyr → Asn	Silent

[a]From ref. 5.

H-2K^{bm3} (positions 77 and 89), have substitutions on the α-helices which locally affect serological sites (9) as well as point into the groove and affect peptide presentation (10).

C57BL/6 (B6, H-2^b) mice are able to mount an ovalbumin (OVA)-specific cytotoxic T lymphocyte (CTL) response when they are immunized with syngeneic cells transfected with an OVA vector or with spleen cells carrying OVA (13). All the CTL activity is restricted by the H-2K^b class I molecule and specific for the OVA$_{253-276}$ region. The response is quite heterogeneous, since analysis of clones and long-term lines has shown that T cell receptor V$_\beta$5, V$_\beta$8, and other V$_\beta$'s are represented in the CTL response. In an H-2^b animal, K^b can therefore positively select OVA-specific T cells.

By analyzing the ability of variant H-2K^b molecules to select positively T cells that respond to H-2K^b plus OVA, or vesicular stomatitis virus nucleoprotein (VSV N), we provide evidence that self-peptides, presented in the groove of the class I molecule on thymus epithelial cells, are critically involved in positive selection of the T cell repertoire. Furthermore, the ability of four different H-2K^b variants to select these responses in the thymus correlates with their ability to present the OVA and VSV N peptides, suggesting the possibility that a self-peptide mimic of the foreign peptide may be involved in positive selection.

RESULTS AND DISCUSSION

In order to study the influence of changes in the class I molecule on the selection of the K^b-restricted repertoire, bone marrow cells from (B6.PL-Thy1^a × K^b-mutant) F$_1$ mice were injected into lethally irradiated B6 or K^b mutant mice. After 8–12 weeks, when bone marrow–derived T cells had matured in the irradiated hosts, the mice were primed with OVA on H-2^b cells. Spleen cells from these mice were boosted *in vitro* with E.G7-OVA (an EL4 transfectant synthesizing OVA) or with allogeneic H-2^d spleen cells and CTL activity was measured 5 days later. The H-2^d–specific effector cells were typed for their expression of Thy1.1 (indicating bone marrow origin) or Thy1.2 only (indicating host origin). The results in Table II show that cells which matured in an irradiated K^{bm1} or K^{bm8} host were not able to respond to K^b plus OVA. On the other hand, T cells which matured in a wild-type K^b, K^{bm3}, and K^{bm5} host did respond to OVA measured as the specific lysis of E.G7-OVA targets. Spleen cells from all groups could respond to H-2^d cells in primary mixed lymphocyte culture to generate CTLs which lyse P815 (H-2^d) target cells, and 70–92% of this alloreactive CTL activity was sensitive to anti-Thy1.1 antibody and complement and therefore due to CTL of bone marrow origin. The failure of the [B6.PL × bm8 → bm8] chimeras to respond to OVA implies strongly that peptide is important in repertoire selection.

TABLE II
Ability of K^b-Mutant MHC Molecules to Select the CTL Response to OVA Assayed in [$F_1 \rightarrow$ Parent] Radiation Chimeras[a]

	OVA-specific response		Alloreactive response		
	% Specific lysis of targets		% Specific lysis of targets		% Donor[b] T cells
Radiation chimeras	EL4	E.G7	EL4	P815	
Experiment 1					
[(B6.PL × bm1)$F_1 \rightarrow$ bm1]	3/0	0/1	4/2	48/30	92
[(B6.PL × bm3)$F_1 \rightarrow$ bm3]	10/6	41/27	2/0	41/18	91
[(B6.PL × bm5)$F_1 \rightarrow$ bm5]	8/4	37/27	7/4	47/24	83
[(B6.PL × bm8)$F_1 \rightarrow$ bm8]	6/3	10/10	6/3	43/24	74
Experiment 2					
[(B6.PL × bm5)$F_1 \rightarrow$ B6]	8/3	56/47	3/4	70/49	83
[(B6.PL × bm5)$F_1 \rightarrow$ bm5]	7/4	56/42	6/3	77/58	85
[(B6.PL × bm8)$F_1 \rightarrow$ B6]	2/3	58/54	4/2	65/47	78
[(B6.PL × bm8)$F_1 \rightarrow$ bm8]	10/6	7/8	6/4	69/54	70

[a]Radiation chimeras were made by injecting 10^7 F_1 bone marrow cells depleted of T cells into 950 rad lethally irradiated recipients of the indicated type, [F_1 bone marrow $\rightarrow$ irradiated host] (1,2). The chimeras were maintained on antibiotic water and were primed after 8–12 weeks by intravenous injection of 2.5×10^7 irradiated spleen cells containing OVA as described previously (14). In experiment 1 spleen cells syngeneic with the marrow donor were used to prime, and in experiment 2 B6.PL spleen cells were used. Seven days after priming, spleen cells from the chimeras were stimulated in culture with irradiated (12,000 R) E.G7-OVA cells (13) to boost the OVA-specific response or with 3000-R-irradiated BALB/c (H-2^d) spleen cells to induce an alloreactive response. After 5 days in culture CTL activity was assayed in a 3-hr ^{51}Cr release assay. Percent specific lysis at two effector/target ratios are shown: 6:1 and 2:1 in experiment 1 and 100:1 and 33:1 in experiment 2.

[b]CTL activity from cultures stimulated with allogeneic BALB/c (H-2d) spleen cells assayed on P815 (H-2^d) targets was typed for sensitivity to anti-Thy1.1 antibody (19E12) (15) and anti-Thy1.2 antibody (13.4) (16) plus complement as described to determine the fraction of donor-derived (Thy1.1 × Thy1.2) effector cells. Briefly, effector cells were treated with complement alone or with one or both antibodies plus complement for 45 min at 37°C before assaying on ^{51}Cr-P815 target cells. Controls with Thy1.1 and Thy1.2 CTL from B6.PL (Thy1.1) and B6 (Thy1.2) mice showed that both antibodies were Thy1 allele specific in their activity.

The nonresponsiveness of the [B6.PL × bm8 $\rightarrow$ bm8] and [B6.PL × bm1 $\rightarrow$ bm1] chimeras to K^b plus OVA could possibly be due to tolerance to the mutant K^b molecule. This is unlikely since, as shown in Table III, (B6.PL × bm8)F_1 and (B6.PL × bm1)F_1 mice do respond to OVA plus K^b. However, the homozygous mutant mice B6.C-H-2^{bm1} and B6.C-H-2^{bm8} are themselves nonresponders to

TABLE III
Failure to Respond to OVA in the [F$_1$ → Parent]
Radiation Chimeras Is Not Due to Tolerance[a]

Responder	% Specific lysis of targets	
	E.G7-OVA	EL4
Experiment 1		
B6	60/41	17/9
(B6 × bm1)F$_1$	47/30	18/7
(B6 × bm3)F$_1$	66/53	25/8
(B6 × bm5)F$_1$	58/46	14/6
(B6 × bm8)F$_1$	53/36	9/8
Experiment 2		
B6	60/53	0/0
bm1	4/2	2/2
bm3	41/25	3/0
bm5	54/39	3/1
bm8	12/6	23/15

[a]The indicated responder mice were immunized by intravenous injection of 2.5×10^7 1000-rad-irradiated syngeneic spleen cells that had been "loaded" with OVA (13). Seven days later spleen cells from the primed animals were restimulated in culture with syngeneic spleen cells that had been incubated with a CNBr digest of OVA, 100 μg/ml, and washed. The CNBr fragments of OVA include the K^b-restricted targeting peptide. CTL activity was measured 5 days later and is expressed as % specific lysis at effector/target ratios of 50:1 and 17:1. The ability of the CTLs to lyse E.G7-OVA targets correlated with their ability to lyse syngeneic K^b-mutant targets in the presence of the OVA peptide.

OVA. The other mutant mice, B6.C-H-2^{bm3} and B6.C-H-2^{bm5}, mount a CTL response to OVA which cross-reacts on OVA plus K^b (Table III). Thus we see a correlation between the ability of four K^b mutants to make a CTL response to OVA and their ability to select the OVA-plus-K^b repertoire in radiation chimeras. To study this more closely we analyzed the ability of the K^b variants to present the K^b-restricted OVA peptide *in vitro*. Several heterogeneous CTL lines from B6 mice specific for K^b plus OVA$_{253-276}$ were tested for lysis of target cells expressing different class I molecules, and target cells expressing K^{bm1} or K^{bm8} were in no case recognized in the presence of peptide (17,18). In addition, none of the 11 clones from B6 mice specific for K^b plus OVA$_{253-276}$ could react with OVA plus K^{bm1} or K^{bm8} (18 and J. Nikolić-Žugić, unpublished results).

The inability of cells derived from K^{bm1} or K^{bm8} mice to present the OVA peptide suggests a trivial explanation for the failure of such mice to select the response in the $[F_1 \rightarrow parent]$ chimeras. Thus, although we typed the alloreactive CTLs as >70% of bone marrow donor origin, it could be that the cells in the chimeras which present OVA to resting CTLs are still of host origin. If this were the case, then [B6.PL × bml → bml] and [B6.PL × bm8 → bm8] chimeras would have inappropriate presenting cells for the K^b-plus-OVA response. This explanation is unlikely to be true for two reasons. First, previous work suggests that antigen-presenting cells turn over rapidly following irradiation (19), and second, we always injected the mice with OVA-bearing immunogenic cells which expressed H-2K^b. However, we did test the possibility by assaying the ability of spleen cells from these chimeras to stimulate anti–H-2K^b CTLs. The data in Table IV show that splenocytes from [B6.PL × bml → bml] and [B6.PL × bm8 → bm8] radiation chimeras are as effective in stimulating a K^b-specific CTL response as are cells from normal F_1 animals. We conclude that the nonresponding chimeras do have appropriate H-2K^b–expressing presenting cells.

The H-2K^{bm8} molecule differs from wild-type H-2K^b in four positions in a β-strand in the floor of the peptide binding site. Two of these, at positions 22 and 24, point up into the site, 23 points down, and 30 is located away from the peptide or TCR contact area and is therefore considered silent (5). The inability of K^{bm8} to select the rather heterogeneous response to K^b plus OVA$_{253-276}$ suggests

TABLE IV
$[F_1 \rightarrow Parent]$ Radiation Chimeras Contain K^b-Expressing Antigen Presenting Cells[a]

Responder spleen cells	Stimulator spleen cells	Anti-K^b specific response (% specific ^{51}Cr release)	
		P815	EL4
Experiment 1			
bm1	B6	13/4	58/37
	(B6.PL × bm1)F$_1$	16/6	64/48
	[(B6.PL × bm1)F$_1$ → bm1]	12/8	55/39
bm8	B6	10/1	67/62
	(B6.PL × bm8)F$_1$	8/1	64/47
	[(B6.PL × bm8)F$_1$ → bm8]	8/3	67/47

[a]Responder spleen cells from naive B6.C-H-2^{bm1} (bm1) or B6.C-H-2^{bm8} (bm8) mice were cultured *in vitro* for 5 days with an equal number of 3000-rad-irradiated spleen cells from stimulator mice as shown. CTL activity was assayed 5 days later on P815 (H-2^d) and EL4 (H-2^b) target cells at effector/target ratios of 48:1 and 16:1.

very strongly that self-peptides associated with the class I molecules of thymic epithelial cells play a critical role in selecting the T cell repertoire. The idea that tissue-specific self-peptides (20,21) or erroneous self-peptides (22) are involved in positive selection has been suggested previously, and in some cases background genes (i.e., not the restriction element) have been shown to influence selection (23–25).

The alternative explanation for our results is that one or more of the K^{bm8} mutations act by altering the architecture of the α-helices, thereby influencing the MHC class I contact with TCR. Although X-ray crystallography is the only technique that can directly answer this issue, available serologic, structural, and functional (see below) results argue strongly against this explanation. The fact that for four H-2K^{bm} molecules the ability to select positively an OVA-plus-K^b–specific T cell repertoire in the thymus correlates with their ability to present the OVA peptide indicates that self-peptides use some form of molecular mimicry to select positively the T cell repertoire for foreign peptides.

The mechanism for this mimicry is suggested by the studies of Garrett et al. (12) describing the three-dimensional structure of a second human class I protein, HLA-Aw68. Comparison of the HLA-A2 and HLA-Aw68 structures revealed that MHC polymorphism apparently alters subsites or pockets within the binding cleft while leaving the overall architecture of the molecule unchanged. Such changes clearly alter peptide binding. A unique set of peptides was bound to the HLA-Aw68 molecule, since the uninterpreted electron density in the cleft differed from that found in HLA-A2 (12). The specific requirements of the individual subsites within the cleft give rise to the observed differences in peptide binding.

Modeling of H-2K^b and H-2K^{bm8} according to the HLA-A structure predicts that the mutations in H-2K^{bm8} are likely to alter an important antigen-contacting pocket. Another implication from this modeling is that other parts of the H-2K^{bm8} peptide-binding groove should be unperturbed, i.e., identical to H-2K^b. Using the H-2K^b–restricted peptide derived from the VSV N protein (S. G. Nathenson, personal communication), we could demonstrate that the rest of the H-2K^{bm8} peptide binding groove is similar to the wild type and that it is functionally equivalent to H-2K^b, because H-2K^{bm8} can efficiently present VSV N to H-2K^b–and H-2K^{bm8}–restricted T cells (J. Nikolić-Žugić, unpublished results and Table V). Based on these results, we speculated that the defect in OVA presentation and repertoire selection by H-2K^{bm8} is due to the mutation which renders the critical peptide-binding pocket nonpermissive for correct binding of OVA-mimicking self-peptide(s) and OVA peptide. Although the binding of self-peptides is difficult to study, we could study the interaction of H-2K^{bm8} and OVA peptide. We raised CTLs against VSV N and examined the ability of the OVA peptide to inhibit presentation of VSV N by H-2K^b or H-2K^{bm8}. An influenza nucleoprotein peptide, NP$_{365-380}$, which binds to D^b was used as a control. $OVA_{253-276}$ could readily inhibit the interaction of VSV N and H-2K^b but had no effect on the VSV N : H-2K^{bm8} interaction (Table V); yet a short minimal determinant $(OVA_{257-264})$ binds to H-2K^{bm8} but still cannot be presented to T cells (J.N.Ž., unpublished). The results indicate that H-2K^{bm8} is not capable of binding the OVA peptide, in the conformation recognized by T cells.

If the theory of molecular mimicry is correct, one would expect the H-2K^{bm8} molecule not only to present VSV N but to also select positively the T cell repertoire specific for it. Table VI shows that this is indeed the case, as [F$_1$ → H-2K^{bm8}] chimeric animals can mount a strong anti–VSV N CTL response that is restricted to H-2K^b.

Therefore, the H-2K^b–restricted antigens, OVA and VSV N, bind to different pockets in the peptide binding site. Both peptides occupy most of the site, since

TABLE V
Inhibition of VSV N Presentation by $OVA_{253-276}$: K^b but Not K^{bm8} Can Bind $OVA_{253-276}$

Presenting molecule	Competitor	Molar excess	% Relative ^{51}Cr release[a]
H-2K^b	NP$_{365-380}$	360	95.2
		120	99.8
		40	102.1
		20	97.2
	OVA$_{253-276}$	360	22.7
		120	56.4
		40	87.9
		20	98.3
H-2K^{bm8}	NP$_{253-380}$	360	94.9
		120	101.9
		40	100.1
		20	99.9
	OVA$_{253-276}$	360	99.2
		120	98.7
		40	103.4
		20	99.4

[a]An effector CTL line was derived from H-2K^{bm8} mice after intraperitoneal immunization with 10^6 plaque-forming units (pfu) VSV. Target cells expressing H-2K^b or H-2K^{bm8} (18) were preincubated with indicated concentrations of inhibiting peptides for 15 minutes before the addition of 1 μM VSV N peptide. After 1 hr of incubation, ^{51}Cr-labeled targets were washed and incubated for 3 hr with anti-VSV CTL line at a 10:1 effector-to-target ratio. The results are expressed as: % relative ^{51}Cr release = (% specific ^{51}Cr release in the presence of competitor) (% specific ^{51}Cr release without competitor). Actual ^{51}Cr release without inhibitor was between 45 and 56%.

TABLE VI
Response of $F_1 \to$ P Chimeras to VSV N Protein[a]

	% Specific lysis	
Radiation chimeras	EL4	N1[b]
[(B6.PL × bm5)$F_1 \to$ bm5]	2/1	70/64
[(B6.PL × bm5)$F_1 \to$ B6]	1/1	69/59
[(B6.PL × bm8)$F_1 \to$ bm8]	2/1	72/62
[(B6.PL × bm8)$F_1 \to$ B6]	0/2	67/64

[a]Experiments were performed exactly as described in Table III except for immunization of chimeras (10^6 pfu VSV i.p.) and *in vitro* stimulation (25×10^6 syngeneic F_1 spleen cells coated with VSV N peptide).

[b]N1 cells [N protein transfected EL4 line (26)] were used as targets in the ^{51}Cr release assay.

they can compete with each other for binding to H-2K^b. However, critical contact for these peptides is provided by different binding pockets. The same pockets are used for OVA- and VSV-mimicking self-peptides during the positive selection of the T cell repertoire specific for these antigens. The distortion of an OVA-binding pocket in H-2K^{bm8} renders this molecule incapable of OVA presentation and anti-OVA T cell repertoire selection. The fact that the VSV N pocket is conserved and functional in H-2K^{bm8} allows for presentation of and repertoire selection for this antigen. This also demonstrates very local effects of the H-2K^{bm8} mutation, rendering the change of helical structure(s) by an allosteric effect even less likely.

The correlation between antigen presentation and repertoire selection by a given MHC molecule may not be strict for all antigens (27). Which peptides govern positive selection and how the interplay between positive and negative selection allows the emergence of a broad and nonautoaggressive T cell repertoire are discussed elsewhere (28,29). As other defined H-2K^b–restricted antigenic peptides become available, we should be able to determine how general are the principles which apply for the antigens studied here.

SUMMARY

Using H-2K^b mutant molecules and the H-2K^b–restricted response to OVA and VSV N peptides, we demonstrate that the positive selection of a T cell repertoire in the thymus requires the self-peptide to be present in the MHC antigen binding groove. We observed a correlation between the ability of four

MHC molecules to present antigenic peptide and their ability to select positively T cells specific for it. The self-peptide involved in positive selection may therefore mimic the foreign peptide during intrathymic selection by binding to the same antigen-binding pocket of MHC peptide-binding site.

REFERENCES

1. Bevan, M. J. (1977) *Nature* **269,** 417–418.
2. Zinkernagel, R. M., *et al.* (1978) *J. Exp. Med.* **147,** 882–896.
3. Kisielow, P., *et al.* (1988) *Nature* **335,** 730–733.
4. Sha, W. C., *et al.* (1988) *Nature* **336,** 73–76.
5. Bjorkman, P. J., *et al.* (1987) *Nature* **329,** 506–512.
6. Davis, M. M., and Bjorkman, P. J. (1988) *Nature* **334,** 395–402.
7. Claverie, J.-M., Prochnicka-Chalufour, A., and Bougueleret, L. (1989) *Immunol. Today* **10,** 10–14.
8. Chothia, C., Boswell, D. R., and Lesk, A. M. (1988) *EMBO J.* **7,** 3745–3755.
9. Nathenson, S. G., *et al.* (1986) *Annu. Rev. Immunol.* **4,** 471–502.
10. Bjorkman, P. J., *et al.* (1987) *Nature* **329,** 512–518.
11. Bluestone, J. A., *et al.* (1986) *J. Immunol.* **137,** 1244–1250.
12. Garrett, T. P. J. *et al.* (1989) *Nature* **342,** 692–696.
13. Moore, M. W., Carbone, F. R., and Bevan, M. J. (1988) *Cell* **54,** 777–785.
14. Carbone, F. R., and Bevan, M. J. (1990) *J. Exp. Med.* **171,** 377–387.
15. Bernstein, I. D., Tam, M. R., and Nowinski, R. C. (1980) *Science* **207,** 68–71.
16. Marshak-Rothstein, A., *et al.* (1979) *J. Immunol.* **122,** 2491–2497.
17. Nikolić-Žugić, J., and Bevan, M. J. (1990) *Nature* **344,** 65–67.
18. Nikolić-Žugić, J., and Carbone, F. R. (1990) *Eur. J. Immunol.* **20,** 2431–2437.
19. Ron, Y., Lo, D., and Sprent, J. (1986) *J. Immunol.* **137,** 1764–1771.
20. Marrack, P., and Kappler, J. (1988) *Immunol. Today.* **9,** 308–315.
21. Marrack, P., McCormack, J., and Kappler, J. (1989) *Nature* **338,** 503–505.
22. Kourilsky, P., and Claverie, J.-M. (1989) *Cell* **56,** 327–329.
23. Ogata, M. *et al.* (1986) *J. Immunol.* **136,** 1178–1185.
24. Singer, A., *et al. Prog. Immunol.* **6,** 60–66.
25. Fry, A. M., Cotterman, M. M., and Matis, L. A. (1989) *J. Immunol.* **143,** 2723–2729.
26. Puddington, L., Bevan, M. J., Rose, J. K., and Lefrancois, L. (1986) *J. Virol.* **60,** 708.
27. Kast, W. M., deWaal, L. P., and Meleif, C. J. M. (1984) *J. Exp. Med.* **160,** 1752–1766.
28. Nikolić-Žugić, J., and Carbone, F. R. (1991) *The Year in Immunology* **10,** 54–65.
29. Nikolić-Žugić, J. (1991) *Immunol. Today* **12,** 65–70.

10

Mechanisms of Peripheral Tolerance

GÜNTER J. HÄMMERLING[1], GÜNTHER SCHÖNRICH[1],
FRANK MOMBURG[1], MARIE MALISSEN[2],
ANNE-MARIE SCHMITT-VERHULST[2],
BERNARD MALISSEN[2], AND BERND ARNOLD[1]

1. Division of Somatic Genetics, Tumor Immunology Program
German Cancer Research Center
6900 Heidelberg, Germany
and
2. INSERM
Centre d'Immunologie de Marseille-Luminy
13288 Marseille, France

INTRODUCTION

A central question in immunology is how the immune system differentiates between self and nonself and thus avoids autoimmunity. In 1959 Burnet (1) and Lederberg (2) developed the hypothesis postulating physical deletion of self-reactive lymphocytes as a mechanism for the acquisition of self-tolerance. This concept of clonal deletion was only recently experimentally verified with the observation that in I-E–positive strains $V_\beta 17$-positive cells were deleted in the thymus (3) and that in T cell receptor (TCR) transgenic mice deletion on $CD4^+8^+$ double positive immature thymocytes was observed when they met the relevant self-antigen (MHC + x) in the thymus (4–6). These and other studies firmly established the existence of clonal deletion in the thymus.

Besides the induction of tolerance in the thymus there is also a need for establishment of tolerance in the periphery because (1) there are probably many antigens in the periphery which are not expressed in the thymus (e.g., tissue-specific antigens) and which may not be transported to the thymus, and (2) deletion in the thymus is a stochastic event which probably does not eliminate all of the autoreactive T cells.

Molecular Mechanisms of
Immunological Self-Recognition

Indeed, several studies have demonstrated the existence of peripheral tolerance induction (7–11). Here we summarize our own findings on the establishment of peripheral tolerance (12, 13).

RESULTS

The Experimental System: Double Transgenic Mice

By generating two types of transgenic mice we have approached the questions of whether and how extrathymic events can contribute to self-tolerance. For this purpose the major histocompatibility complex (MHC) class I gene K^b was expressed outside the thymus in a tissue-specific manner on an allogeneic background. A class I molecule was chosen as a model antigen because the T cell receptor recognizes intact MHC molecules but not degradation products which could migrate to the thymus. (In this case the MHC fragments would be recognized only as peptides presented by a thymic MHC molecule.) In addition, in previous studies it was established in transgenic mice that soluble class I molecules did not induce T cell tolerance for the respective cell-bound class I molecule (14). We used the albumin promoter and the keratin IV promoter for liver-specific and epithelium-specific expression of K^b and the glial fibrillary acidic protein (GFAP) promoter to restrict expression of K^b to cells of neuroectodermal origin. Expression of the transgene was verified on the mRNA level by cDNA synthesis and polymerase chain reaction (PCR) analysis using class I allele-specific oligonucleotides and on the protein level by immunohistology. In all three cases transgenic mice were obtained which expressed the K^b antigen exclusively outside the thymus: either only in hepatocytes (albumin promoter), or only in epithelial cells of hair follicles in the epidermis (keratin VI promoter), or only in cell types of neuroectodermal origin (GFAP).

Second, genes coding for a K^b-specific TCR were introduced into the germline of mice (12). The respective $\alpha\beta$ TCR genes were derived from a K^b-specific and strictly CD8-dependent cytotoxic T lymphocyte (CTL) clone isolated from B10.BR ($H\text{-}2^k$) mice (15). These TCR transgenic mice were crossed with each of the above-mentioned K^b transgenic mice. All these mice were completely tolerant *in vivo* to K^b because they failed to reject K^b-positive skin and K^b-positive tumors. An example is shown in Table I. No evidence for autoimmune reactions (lymphocyte infiltration) was observed. However, *in vitro* a cytotoxic response was obtained after stimulation with K^b-positive spleen cells, demonstrating that the observed *in vivo* tolerance could easily be broken under *in vitro* conditions, where various cytokines are probably produced.

TABLE I
Tolerance in GFAP-K^b and TCR × GFAP-K^b Mice toward the K^b Antigen

Mice expressing			Skin graft survival^a	EL-4 tumor growth^b		
TCR	GFAP-K^b	haplotype	>60 days	1×10^6	3×10^5	5×10^4
−	−	(dxbm1)	0/8	+++	+++	+++
−	+	(dxbm1)	9/0			
−	−	(dxk)		+++	−	−
+	−	(dxk)		+	−	−
−	+	(dxk)		+++	+++	+++
+	+	(dxk)		+++	+++	+++

Source: Ref. 13.

^{a}Mice were crossed with bm1 mice and grafted with C57Bl/6 skin. Number of mice with skin accepted over number of mice grafted after 60 or more days.

^{b}The indicated numbers of EL-4 cells were administered subcutaneously to 5–7 mice of each type. They were screened for tumor growth after 10–14 days.

Lack of Tolerance Induction in the Thymus

Although the observed lack of thymic expression of the K^b transgene controlled by the various tissue-specific promoters strongly argued against tolerance induction in the thymus in these mice, this had to be formally proved. The availability of anticlonotypic antibodies against the TCR allowed us to follow the fate of the K^b-reactive T cells in the double transgenic mice. Clonotype$^+$, CD8$^+$CD4$^-$ mature T cells could be detected in normal numbers in the thymus of all of these double-transgenic mice, demonstrating that there was no clonal deletion in the thymus (12, 13). These mature thymocytes could proliferate in response to anticlonotypic antibodies and developed into K^b-specific CTLs *in vitro* after stimulation with antigen in the absence of exogenous interleukin-2 (IL-2). The response was identical to that of thymocytes from TCR single transgenic mice. Hence, it can be concluded that the observed tolerance to K^b was caused by an extrathymic event. Even if one considers the possibility that undetectable amounts of H-2K^b were present in the thymus of the different transgenic mice, they clearly had no influence on the maturation of precursors into anti-K^b cells in the thymic microenvironment.

DISTINCT MECHANISMS OF PERIPHERAL TOLERANCE

The search for the presence of clonotype[+] T cells in the periphery (spleen and lymph nodes) of the double transgenic mice revealed the existence of distinct mechanisms of peripheral tolerance depending on the tissue-specific expression of the K^b transgene, as described below.

GFAP-K^b Mice

Contact with the K^b antigen in the periphery led to a strong reduction of clonotype[+], CD8[+] T cells in spleen and lymph nodes of (TCR × GFAP-K^b)F$_1$ mice in comparison to the numbers found in TCR single-transgenic mice (Table II). This reduction could be due to either physical elimination of these cells or modulation of both TCR and CD8 molecules on the cell surface. In order to distinguish between deletion of lymphocytes and down-regulation of TCR on the cell surface, we analyzed spleen cells from TCR single- and (GFAP-K^b × TCR)F$_1$ double-transgenic mice for expression of Thy-1 and CD3 molecules. If the hypothesis of down-regulation of TCR molecules was correct, the disappearance of clonotype[+] CD8[+] cells should be accompanied by increased appearance of Thy-1[+], CD3[−] lymphocytes in double- versus single-transgenic mice. Indeed, in the (GFAP.K^b × TCR)F$_1$ mice a very high number of Thy-1[+] CD3[−] cells was found: 75% in comparison to 21% in TCR single transgenic mice (see Table II). These data strongly support the conclusion that down-regulation of the TCR had occurred in these double-transgenic mice.

Next, we investigated whether the observed down-regulation of the TCR was reversible. For this purpose clonotype[−] cells were isolated from the tolerant (GFAP.K^b × TCR)F$_1$ mice by using a Fluorescene-activated Cell Sorter (FACS). Coculture of the clonotype[−] cells with K^b-expressing C57Bl/6 spleen cells in a 4-day mixed lymphocyte culture (MLC) resulted in clonotype[+] CD8[+], K^b-reactive CTLs. This up-regulation of TCR and CD8 could be detected only on clonotype-negative T cells from (TCR × GFAP-K^b)F$_1$ mice, not on clonotype-negative T cells sorted from TCR single-transgenic animals. When the number of clonotype[−] T cells was determined in the spleen using either Thy-1 or CD2 as a pan T cell marker, a surprisingly high percentage (75%) of such T cells was found in the (GFAP.K^b × TCR)F$_1$ mice, which clearly contrasted with the 21% found in TCR single transgenics (Table II). These 75% probably represent the T cells with the down-regulated clonotype. Thus, the observed reduction of clonotype[+] CD8[+] T cells in the periphery of (TCR × GFAP-K^b)F$_1$ mice is, at least in part, based on reversible down-regulation of TCR and CD8 molecules. The mechanism by which the clonotype[−] cells can be activated *in vitro* by K^b-positive C57Bl/6 spleen cells is not clear. These C57Bl/6 spleen

TABLE II
Tolerance to Extrathymic K^b Antigen Depending on Cell Type Expression^a

Parameter	TCR$^+$	Ker IV-K^b × TCR	GFAP-K^b × TCR	Alb-K^b × TCR
% clono$^+$, CD8$^+$ cells in spleen	12 ± 3	12 ± 3	2 ± 2	1.0 ± 0.5
% Thy-1$^+$CD3$^-$ cells	21	17	75	67
Conclusion: Mechanism of tolerance induction	—	Nonrespon-siveness	Down-regulation of TCR and CD8, reversible by MLR	Down-regulation TCR and CD8, reversible by stimulation through alternative pathway

Source: Ref. 13.

^{a}Extrathymic tissue expression of the K^b transgene under the control of tissue-specific promoters: GFAP-K^b (glial fibrillary acidic protein) on astrocytes, choroid plexus, ependymal cells of the brain, and Schwann cells of the small intestine; keratin IV-K^b on certain keratinocytes; albumin-K^b on hepatocytes. All animals are of H-2$^{d×k}$ haplotype. Ig$^-$, CD4$^-$, Thy-1$^+$ splenocytes were analyzed by FACS for the expression of CD3. Note the drastic increase of TCR-negative (CD3$^-$) T cells in GFAP.K^b and Alb.K^b mice, suggesting down-regulation of the clonotype as a mechanism of tolerance.

cells differ from the transgenic responders also in MHC class II and the D end. Therefore, the third-party reaction against I-A^b or H-2D^b could result in the release of cytokines which could induce reappearance of the TCR on the downregulated cells.

Albumin-K^b Mice

A different mechanism for the establishment of peripheral tolerance appears to be operative in (Alb.K^b × TCR)F$_1$ mice which express the K^b exclusively on hepatocytes owing to the use of the albumin promoter. Again, a strong reduction of clonotype$^+$, CD8$^+$ cells could be observed in spleen and lymph nodes in comparison to the numbers obtained in TCR single-transgenic mice. However, when the clonotype$^-$ cells were isolated by FACS, no up-regulation of TCR molecules *in vitro* could be detected after stimulation with C57Bl/6 spleen cells as was found for the (TCR × GFAP-K^b)F$_1$ mice. Since in these mice there was also a very high number of Thy-1$^+$CD3$^-$ cells (Table I), peripheral deletion of the clonotype$^+$CD8$^+$ T cells is extremely unlikely. It rather appears that the

Alb.K^b mice peripheral tolerance is induced by irreversible down-regulation of the TCR, which appears to be more tight than the down-regulation observed in the (GFAP.$K^b \times$ TCR)F_1 mice because in the latter an MLR leads to up-regulation of the clonotype. Preliminary studies suggest that the down-regulated cells from the (Alb.$K^b \times$ TCR)F_1 mice are not completely refractory to activation because stimulation with cross-linked anti-CD2 antibodies could induce expression of the TCR (unpublished).

Keratin IV-K^b Mice

A third mechanism for the induction of extrathymic tolerance was seen in (TCR $\times$ keratin IV-K^b)F_1 animals expressing K^b only on certain keratinocytes outside the thymus. The number of clonotype$^+$ CD8$^+$ cells in spleens and lymph nodes of these mice was unchanged in comparison to TCR single-transgenic mice. Nevertheless, these mice were fully tolerant *in vivo* (no rejection of K^b-positive grafts) and no sign of autoimmunity could be seen. This apparent T cell unresponsiveness was reversible *in vitro*: K^b-specific CTLs were found after stimulation with C57Bl/6 spleen cells in the absence of exogenous IL-2 (Table II).

CONCLUSIONS

The present data demonstrate that peripheral tolerance can be mediated by several distinct mechanisms. These include the existence of clonotype-positive but *in vivo* functionally inert T cells (keratin IV-K^b mice), down-regulation of the TCR which is reversible by MLR (GFAP.K^b mice), and down-regulation of the TCR not reversible by MLR but reversible by stimulation with anti-CD2 (albumin.K^b mice) (see Table I). It is not clear whether these different mechanisms are caused by qualitatively distinct tolerogenic signals provided by the respective tissue (e.g., hepatocytes versus keratiocytes or GFAP-positive cells or by different density of the K^b transgene, different length of time, or multiplicity of contact between the cells expressing the K^b transgene and the T lymphocytes, etc.

Certainly these data do not exclude that additional mechanisms for the establishment and maintenance of tolerance can be operative. However, in the present studies we did not find evidence for physical disappearance (deletion) of peripheral T cells as was described by others (10, 16, 17). But it is possible that some peripheral deletion also takes place and contributes to the peripheral tolerance in the mice described here.

In this context, note that the present data also show that in the same animal we can observe distinct manifestations of peripheral tolerance. For example, in the (TCR × GFAP-K^b)F$_1$ mice, in addition to cells with down-regulated TCR, there exist clonotype$^+$ CD8$^+$ cells, albeit in strongly reduced numbers. However, one should keep in mind that this remaining 1 or 2% of the clonotype$^+$ cells with potential reactivity against K^b is actually a rather high number when compared to a normal immune system.

In conclusion, the studies described here reveal a surprising variety of manifestations of peripheral tolerance. It appears that nature has developed multiple backup systems to take care of the potentially autoreactive cells that have not been physically deleted in the thymus. The observations may be of clinical relevance because distinct mechanisms of extrathymic tolerance are likely to differ in their efficiency in protecting against the breakdown of self-tolerance and therefore against autoimmunity. The finding that different tissues can induce different levels of tolerance could explain why some organs are more susceptible to autoimmune diseases than others.

ACKNOWLEDGMENTS

We thank G. Küblbeck, L. Jatsch, A. Klevenz, and S. Beck for excellent technical assistance, K. Hexel for expertise in running the cell sorter, and U. Esslinger for preparation of the manuscript.

REFERENCES

1. Burnet, F. M. (1959) *The Clonal Selection Theory of Acquired Immunity, Cambridge Univ. Press.*
2. Lederberg, J. (1959) *Science* **129**, 1649.
3. Kappler, J. W., Roehm, N., and Marrack, P. (1987) *Cell* **49**, 273.
4. Kisielow, P., Blüthmann, H., Staerz, U. D., Steinmetz, M., von Boehmer, H. (1988) *Nature* **333**, 742.
5. Sha, W. C., Nelson, C. A., Newberry, R. D., Kranz, D. M., Russel, J. H., and Loh, D. Y. (1988) *Nature* **336**, 229.
6. von Boehmer, H. (1990) *Annu. Rev. Immunol.* **8**, 531.
7. Rammensee, H.-G., Kroschewski, R., and Frangoulis, B. (1989) *Nature* **339**, 541.
8. Jones, L. A., Chin, L. T., Merriam, G. R., Nelson, L. M., and Kruisbeek, A. M. (1990) *J. Exp. Med.* **172**, 1277.
9. Miller, J. F. A. P., Morahan, G., Allison, J., Bhathal, P., and Cox, K. O. (1989) *Immunol. Rev.* **107**, 109.
10. Rocha, B., and von Boehmer, H. (1991) *Science* 251, 1225.
11. Burkly, L. C., Lo, D., and Flavell, R. A. (1990). *Science* **248**, 1364.
12. Schönrich, G., Kalinke, U., Momburg, F., Malissen, M., Schmitt-Verhulst, A.-M., Malissen, B., Hämmerling, G. J., and Arnold, B. (1991) *Cell* **65**, 293.

13. Hämmerling, G. J., Schönrich, G., Momburg, F., Auphan, N., Malissen, M., Malissen, B., Schmitt-Verhulst, A.-M., and Arnold, B. (1991) *Immunol. Rev.* **122,** 47.
14. Arnold, B., Dill, O., Küblbeck, G., Jatsch, L., Simon, M. M., Tucker, J., and Hämmerling, G. J. (1988) *Proc. Natl. Acad. Sci. USA* **85,** 2269.
15. Hue, I., Trucy, J., McCoy, C., Couez, D., Malissen, B., and Malissen, M. (1990). *J. Immunol.* **144,** 4410.
16. Jones, L. A., Chin, L. T., Longo, D. L., and Kruisbeek, A. M. (1990b) *Science* **250,** 1726.
17. Webb, S., Morris, C., and Sprent, J. (1990). *Cell* **63,** 1249.

11

An Analysis of T Cell Receptor–Ligand Interaction Using a Transgenic Antigen Model for T Cell Tolerance and T Cell Receptor Mutagenesis

BARBARA FAZEKAS DE ST. GROTH[1,*],
PHILLIP A. PATTEN[2,†], WILLIAM Y. HO[2], EDWIN P. ROCK[1],
AND MARK M. DAVIS[1]
1. Howard Hughes Medical Institute and
Department of Microbiology and Immunology
Stanford University
Stanford, California 94305
and
2. Department of Microbiology and Immunology
Stanford University
Stanford, California 94305

INTRODUCTION

We have approached questions of $\alpha\beta$ T cell recognition and selection by a number of routes. To analyze the molecular basis of T cell receptor (TCR) recognition of foreign peptide–major histocompatibility complexes (MHC), we have extensively mutagenized putative CDR "loops" of TCR V_α and V_β

*Current address: Centenary Institute of Cancer Medicine and Cell Biology, The University of Sydney, Sydney NSW 2006, Australia.
†Current address: Department of Chemistry, University of California, Berkeley, Berkeley, CA 94720.

Molecular Mechanisms of
Immunological Self-Recognition

sequences in an attempt to transfer specificity. As discussed below, these experiments have thus far failed to transfer antigen/MHC specificity but have succeeded in transferring *Staphylococcus aureus* enterotoxin B (SEB) reactivity. In the long term, we hope to reconstitute T cell recognition in a completely cell-free fashion. To this end, we have successfully expressed a T cell receptor αβ heterodimer in a soluble form, utilizing signal sequences for lipid linkage in place of the putative transmembrane/cytoplasmic sequences. This approach has also worked for I-E^k, the cognate MHC of the mouse TCR mentioned above. Characterization of this soluble class II MHC molecule has revealed that low pH ($\approx$5.0) greatly increases the ability of functional antigen/MHC complexes to form (D. Wettstein *et al.*, in preparation). This suggests an explanation for the disparity between the reported rate of peptide binding to purified MHC molecules *in vitro* and the vastly greater efficacy of processing by antigen-presenting cells.

Finally, we also discuss data relevant to the central theme of this conference, namely T cell tolerance, utilizing a paired T cell epitope/TCR transgenic combination similar to that developed by Goodnow *et al.* (1) for B cells. This gives us the ability to manipulate the concentration of a specific peptide which the animal now considers to be "self."

T CELL RECEPTOR MUTAGENESIS

T cell receptor sequences, particularly V regions, are very similar to immunoglobulins, and the concensus of a number of experts is that they probably fold and pair in much the same way (2, 3). However, the preponderance of their sequence diversity lies in the V(D)J junctional or CDR3-equivalent region, exceeding by many orders of magnitude the diversity of immunoglobulin CDR3's in the absence of somatic mutation (4). In the remainder of the variable region, TCRs have substantially less sequence diversity than immunoglobulins (4). This dichotomy, together with the structure of the MHC class I molecules with their apparent antigen-binding groove (5), has led a number of groups (3, 6) including our own (4) to suggest that CDR3 loops of T cell receptors may be the principal determinants of peptide specificity. In order to test this hypothesis, we have performed two types of mutagenesis experiments involving predicted CDR loops. The first of these involves transferring the CDR3 loops of both α and β chain TCR sequences from T cells which are specific for antigens on the I-E^k restricting element onto α and β chains from a receptor which recognizes pigeon cytochrome *c* on I-E^k. Three different αβ CDR3 pairs were made, transferred into Jurkat cell recipients, and tested for acquisition of the new specificity. In no case was a transfer of specificity achieved, even when the donor and recipient receptors shared the same V$_\alpha$ element. This result indicates that CDR3 sequences are not solely responsible for peptide specificity but rather are dependent on other sequences for their conformation. It is also possible that other sequences within the binding site are in direct contact with antigen, although it is difficult to imagine non-CDR3 sequences playing a major role, as they account for so little of the diversity.

A second line of experimentation was designed to transfer specificity *in toto* by shifting all the classical CDRs of one V_β sequence (CDR1, 2, and 3) onto another V_β, which was then paired with the donor's V_α. The two original receptors were chosen to share germline V_α sequences. Surprisingly, this wholesale transfer of CDR loops did not confer antigen/MHC specificity, suggesting that T cell recognition may not be as simple as antigen–immunoglobulin recognition. Interestingly, sequential transfer of CDRs did result in a clear transfer of SEB reactivity, CDR2 proving more effective than CDR1. This suggests that toxin reactivity may, in at least some cases, be dependent on the same sequences as antigen/MHC recognition. There is some logic to this as TCRs should be uniquely equipped to recognize MHC, and thus such conserved features as may be necessary to make contact with a variety of MHC molecules may provide targets of opportunity for toxins.

A PAIRED ANTIGEN/TCR TRANSGENIC SYSTEM

In order to develop a model for T cell tolerance with a defined antigen, we have constructed a gene fusion in which the sequence encoding the COOH-terminal 24 amino acids of moth cytochrome *c* has been added to the gene for hen egg white lysozyme (HEL), under the control of the metallothionein promoter, utilizing a plasmid kindly provided by C. Goodnow (1). Two transgenic (HELCYT) lines have been made with this construct and express easily detectable (1–10 ng/ml) serum levels of protein, using an anti-HEL enzyme-linked immunosorbent assay. Crossing these HELCYT mice with T cell receptor transgenics specific for the cytochrome *c* epitope together with I-E^k results in a combination of deletion of the antiself specificity in the thymus together with apparent paralysis (presumptive anergy) of the remaining transgene-bearing, CD4$^+$ peripheral T cells (which can constitute up to 10% of the peripheral blood mononuclear cells). Substantial individual-to-individual variation is seen in the extent of deletion, probably due to variation in the level and time of onset of HELCYT production, since the T cell phenotype within each TCR transgenic line is highly reproducible. The effect of Zn^{2+} induction of the metallothionein-driven HELCYT construct (which increases serum levels of HELCYT by five- to tenfold) is to produce a significant shift toward thymic deletion. This shows that removal of self-reactive cells can be driven by increased amounts of antigen, suggesting that this is an avidity-dependent process (since in this case neither the TCR nor the antigen itself is changing).

As is characteristic of a number of negative selection models, this one shows deletion of the antiself specificity at the double positive ($CD4^+8^+$) stage of thymocyte development. In a series of TCR transgenic mouse lines made using the same TCR constructs, we have seen two clearly distinct modes of self-tolerance induction at the double positive stage. Each transgenic mouse line shows only one of the two phenotypes when subjected to a negatively selecting environment. In the first phenotype, transgenic α chain virtually disappears from the surface of double positive thymocytes while β chain expression is relatively unchanged. The number of double positive cells, which now express endogenously rearranged α chains, is preserved. The second phenotype, similar to that reported by von Boehmer *et al.* (7) and Loh *et al.* (8), shows massive deletion of double positive cells, with retention of transgenic $\alpha\beta$ pairs on the surface of the few remaining double positive cells. These two phenotypes are probably integration site or gene dosage effects, with higher levels of TCR mRNA or protein expression causing more profound phenotypes of deletion.

The stage at which self-tolerance first manifests itself phenotypically is the same in our two self-antigen models, i.e., when the transgenic TCRs are together with the HELCYT fusion gene on an $H-2^k$ or $H-2^{k/b}$ background, or when crossed onto an $H-2^s$ background (this particular TCR, 5C.C7, being alloreactive against $I-A^s$). The only apparent difference between the two types of negative selection is that in $H-2^s$ mice, but not in HELCYT mice, there are many more double negative ($CD4^-8^-$) $\alpha\beta$ transgene-bearing T cells in the periphery. This may reflect a difference in affinity of the 5C.C7 receptor for these two antigens, as the cytochrome $c/I-E^k$ reactivity is relatively robust (i.e., CD4 independent at all but low levels of cytochrome) whereas the $I-A^s$ alloreactivity is not. Thus $CD4^-$ alloreactive T cells might escape thymic selection, whereas cytochrome c reactive T cells might not (or may do so less efficiently). The loss of the coreceptor CD4 from otherwise self-reactive peripheral T cells is reminiscent of the down-regulation of CD8 on transgenic anti–H-Y T cells (7).

ACKNOWLEDGMENTS

We wish to thank NIH for grant support. Barbara Fazekas de St. Groth has been supported by the Irvington House Institute and by the Australian National Health and Medical Research Council, Phillip Patten was supported by a training grant in molecular and cellular biology (NIH), William Ho is supported by an MSTP training grant (NIH), and Edwin Rock is supported by a Howard Hughes Medical Institute predoctoral fellowship.

REFERENCES

1. Goodnow, C. C., Crosbie, J., Adelstein, S., Lavoie, T. B., Smith-Gill, S., Brink, R. A., Pritchard-Briscoe, H., Wotherspoon, J. S., Loblay, R. H., Raphael, K., Trent, R. J., and Basten, A. (1988) *Nature* **334,** 676.

2. Novotny, J., Tonegawa, S., Saito, H., Kranz, D. M., and Eisen, H. N. (1986) *Proc. Natl. Acad. Sci. USA* **83,** 742.

3. Chothia, C., Boswell, D. R., and Lesk, A. M. (1988) *EMBO J.* **7,** 3745.

4. Davis, M. M., and Bjorkman, P. J. (1988) *Nature* **334,** 395.

5. Bjorkman, P. J., Saper, M. A., Samraouri, B., Bennett, W. S., Strominger, J. L., and Wiley, D. C. (1987) *Nature* **329,** 506.

6. Claverie, J. M., Prochnicka-Chalufour, A., and Bougueleret, L. (1989) *Immunol. Today* **10,** 10.

7. Kisielow, P., Bluthmann, U. D., Stearz, U. D., Steinmetz, M., and von Boehmer, H. (1988) *Nature* **333,** 742.

8. Sha, W. C., Nelson, C. A., Newberry, R. D., Kranz, D. M., Russell, J. H., and Loh, D. Y. (1988) *Nature* **336,** 73.

12

T Cell Repertoire and Tolerance

ANN M. PULLEN[1], YONGWON CHOI[1],
JOHN W. KAPPLER[1,2], AND PHILIPPA MARRACK[1,2,3]

1. *Howard Hughes Medical Institute*
Division of Basic Immunology
Department of Medicine
National Jewish Hospital for Immunology and Respiratory Medicine
Denver, Colorado 80206
and
2. *Department of Microbiology and Immunology*
and Department of Medicine
University of Colorado Health Sciences Center
Denver, Colorado 80220
and
3. *Department of Biochemistry, Biophysics, and Genetics*
University of Colorado Health Sciences Center
Denver, Colorado 80220

INTRODUCTION

Murine $\alpha\beta$ T cell receptors (TCRs) have five variable segments, V_α and V_β (variable), D_β (diversity), and J_α and J_β (joining). Additional variability is provided by imprecise joining of these gene segments and the inclusion of additional nucleotides at the junctions (for review see ref. 1). All five variable segments contribute to a binding site for complexes of peptide antigen with major histocompatibility complex (MHC) products. Thymocytes bearing receptors which interact with self-MHC during development in the thymus are selected to mature (positive selection) (2,3). Subsequently, it is thought that thymocytes with receptors with high affinity for self are clonally eliminated, and only positively selected cells which escape this negative selection will leave the thymus for the periphery (reviewed in ref. 4).

Recently, monoclonal antibodies with specificity for V_β elements have been generated and these have been used to study the T cell repertoire and the

129

forces which shape it. A number of studies have made use of these reagents to study positive selection (5–7); however, here we will concentrate on their use to study tolerance and negative selection of T cells.

SUPERANTIGENS

Self-Superantigens

Nearly 20 years ago Festenstein documented strong primary mixed lymphocyte reactions (MLRs) between cells from MHC-identical mouse strains (8,9). The products of the minor lymphocyte-stimulating (Mls) loci were the determinants which stimulated these reactions. Some years later it was demonstrated that the MLR was due to T cells responding to Mls determinants on class II–bearing antigen-presenting cells (10,11). Recently it has been shown that the V_β element of the TCR is pivotal for these responses (12–14). This is in contrast to the response to conventional peptide antigen/MHC complexes, when all the variable components of the TCR are involved (15,16).

T cells bearing one or a few V_β elements respond to stimulation by the Mls determinants (Table I). In mice which express Mls/class II complexes, T cells bearing reactive V_β elements are clonally eliminated during development in the thymus (26,12–14). Similar phenomena have been described for T cells bearing TCRs reactive with moieties combined with IE molecules (19–22). We have coined the term "self-superantigens" to describe these entities which combine with class II molecules to stimulate large numbers of T cells bearing particular V_β elements.

TABLE I
V_β Recognition of Self-Superantigens

Self-superantigen	Reactive V_β elements	References
Unknown + IE (B cell specific)	17a	17, 18
Unknown + IE	5, 11, 12	19–22
Mls-1[a] (chromosome 1)	6, 8.1, 9, 7	12, 13, 23, 24
Mls-2[a] (chromosome 4)	3	14, 25
Mls-3[a] (chromosome 16)	3	14, 25

Foreign Superantigens

The foreign superantigens include toxins made by staphylococci, streptococci, and mycoplasma, which cause food poisoning and/or shock in humans and animals (reviewed in ref. 27). Like the self-superantigens, they form complexes with MHC class II molecules and stimulate T cells via the V_β element of the TCR, with little contribution from the other variable elements of the receptor. For example, staphylococcal enterotoxin B (SEB) stimulates T cells bearing V_β's 3, 7, all members of the $V_\beta 8$ family, and $V_\beta 17$ in mouse (28–30) and T cells bearing V_β's 3, 12, 14, 15, 17, and 20 in humans (31,32).

TOLERANCE TO SELF-SUPERANTIGENS SHAPES THE T CELL REPERTOIRE

The availability of monoclonal antibodies specific for V_β elements has facilitated the study of the repertoire of both laboratory inbred strains and natural populations of wild mice. A comparison of the levels of V_β expression for two related strains, CBA/J (Mls-1[a] and Mls-2[a] or Mls -3[a]) and CBA/CaJ (Mls-1[b], 2[b], 3[b]) (Fig. 1), shows how the clonal elimination of thymocytes reactive with self-superantigen/MHC complexes plays an important role in shaping the resulting peripheral repertoire. T cells bearing V_β's 6, 8.1 and 9 are eliminated during thymic development due to reactivity to Mls-1[a], and $V_\beta 3^+$ thymocytes are

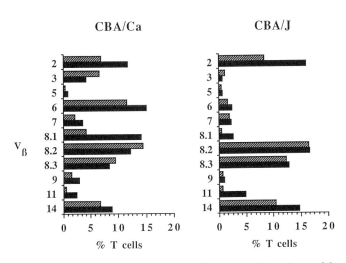

Fig. 1. V_β expression in CBA mice. Peripheral T cells were purified and stained for V_β, CD4, and CD8 expression as previously described (33). Shaded bars indicate CD4$^+$ T cells and solid bars indicate CD8$^+$ T cells.

clonally deleted in response to Mls-2^a or Mls-3^a. Moreover, both strains are of the k haplotype, so they express IE^k and the self-superantigens which combine with this molecule, and therefore these mice have few V$_\beta$5$^+$ or V$_\beta$11$^+$ peripheral T cells.

We analyzed the V$_\beta$ repertoire of natural populations of wild mice trapped around Gainesville, Florida (33). In addition to numerous examples of wild mice which had eliminated T cells with receptors reactive with self-superantigen/MHC complexes, individual mice also had reduced repertoires due to gene deletions or point mutations.

A quarter of the wild mice included in the study were homozygous for an extensive V$_\beta$ gene deletion on chromosome 6. The deletion included 11 V$_\beta$ genes (V$_\beta$5–V$_\beta$15) and so reduced the extent of the repertoire to less than half that of a mouse carrying the full complement of V$_\beta$ genes. All the wild mice analyzed had no V$_\beta$17a$^+$ T cells. They carried a restriction fragment length polymorphism (RFLP) indicative of a point mutation which results in a premature stop codon in the V$_\beta$17b gene (34).

The mouse population as a whole has at least 22 V$_\beta$ genes; however, several of the wild mice analyzed in this study employed all the above strategies to reduce the extent of their T cell repertoire and so had surprisingly limited repertoires. We have recently proposed that mice reduce the extent of their T cell repertoire in order to reduce the likelihood of succumbing to the foreign superantigens (33). These foreign superantigens include a variety of microbial toxins for which T cell stimulation, again primarily via the V$_\beta$ elements, is necessary for their toxicity (35).

The combined effects of V$_\beta$ gene deletions and clonal elimination by self-superantigens such as Mls-1^a allow a great many different V$_\beta$ (and consequently TCR) repertoires to be expressed in any mouse population. This may reflect the balance between the selective advantage for large TCR and V$_\beta$ repertoires, which would allow responses to as many pathogens as possible, and the disadvantage of expansion of certain V$_\beta$'s because of attack by foreign superantigens (enterotoxins).

V$_\beta$ INTERACTION WITH THE SELF-SUPERANTIGEN, Mls-1^a

We are beginning to understand why the V$_\beta$ element is pivotal in the T cell response to the superantigens. The study of the V$_\beta$ repertoire of wild mice mentioned above provided us with some important clues. Among the wild mice there were two variant V$_\beta$8.2 alleles, V$_\beta$8.2b and V$_\beta$8.2c, which had not been seen in laboratory inbred strains (Table II). Interestingly, wild mice expressing Mls-1^a eliminated T cells bearing V$_\beta$8.2c. This was a surprising finding in the light of our previous studies on laboratory strains expressing Mls-1^a, which

TABLE II
Mls-1ᵃ Reactivity of Variant V$_\beta$8.2 Segments

Allele	Origin	% 8.2⁺ T hybrids reactive to Mls-1ᵃ	Amino acid				
			8	22	51	70	71
a	Inbred	10	N	N	G	E	N
b	Wild	50	N	D	G	E	N
c	Wild	100	S	D	D	K	E

showed that T cells bearing V$_\beta$8.2a were not eliminated (12). The generation of T cell hybridomas expressing V$_\beta$8.2b and V$_\beta$8.2c confirmed that all V$_\beta$8.2c⁺ T cells were Mls-1ᵃ reactive and showed that half of the V$_\beta$8.2b⁺ hybrids were Mls-1ᵃ reactive (Table II).

We used site-directed mutagenesis and transfection to express mutant T cell receptors in which the V$_\beta$8.2a allele was converted to the V$_\beta$8.2c allele residue by residue (36). A summary of the data for a panel of transfectants is shown (Table III).

The mutations 22N → D, 70K → E, and 71N → E were important in conferring Mls-1ᵃ reactivity on an otherwise non–Mls-reactive T cell hybridoma, DO-11.10. This hybridoma reacts with an ovalbumin peptide presented in the groove of MHC class II molecules, I-Aᵈ and I-Aᵇ. The mutations conferring Mls reactivity did not affect the response of the hybridomas to this conventional peptide antigen (36).

We have modeled the structure of the DO-11.10 TCR based on immunoglobulin according to Chothia, *et al.* (37). These data are consistent with Mls-1ᵃ interacting with a site on the V$_\beta$ element well away from the complementarity-determining regions of the TCR, which are predicted to interact with the complex of conventional peptide antigen and MHC.

V$_\beta$ INTERACTION WITH THE FOREIGN SUPERANTIGENS

Once we had highlighted this region of the TCR V$_\beta$ as being involved in the interaction with the self-superantigen, Mls-1ᵃ, we wanted to determine whether this site was also involved in the interaction with foreign superantigens. We tested the transfectants expressing the mutated DO-11.10 TCRs for their ability to respond to a panel of staphylococcal enterotoxins. However, we saw no differences in their response patterns, so we turned to an alternative

TABLE III
Summary of Mls-1ᵃ Reactivity of Transfectants Expressing Mutant $V_\beta 8.2$ Elements

	Transfection recipient	
Mutation	DO-11.10.3 $(V_\alpha 13, V_\alpha 1)$	DO-11.10.7 $(V_\alpha 1)$
$8N \to S$	−	−
$22N \to D$	+	−
$51G \to D$	−	−
$70E \to K/71N \to E$	+ +	+
$22N \to D/70E \to K/71N \to E$	+ + +	+ +

system to determine which residues are important in human V_β elements for their reactivity to these toxins.

We demonstrated that human T cells bearing the closely related V_β's 13.1 and 13.2 differed in their ability to respond to the staphylococcal enterotoxins C2 and C3 (SEC2 and SEC3). Sequence comparisons showed that there were several differences between hV_β 13.1 and 13.2 between residues 67 and 78. We generated a transfectant expressing a receptor including a variant $hV_\beta 13.1$ in which residues 67–78 were replaced by the $V_\beta 13.2$ sequence at those positions (38). This transfectant ($V_\beta 13.1/.2$) gained reactivity to SEC2 and SEC3 as shown in Fig. 2. Thus, some of the residues which differ between 13.1 and 13.2 (67–78) must be important for binding of these foreign superantigens.

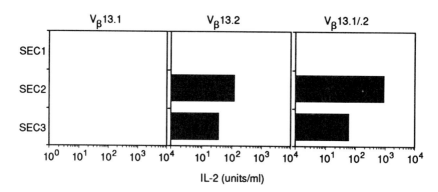

Fig. 2. Residues 67–78 of human V_β affect toxin reactivity. Transfectants expressing chimeric murine/human TCRs (38) were tested for their response to staphylococcal enterotoxins SEC1, SEC2, and SEC3 presented by 10^5 A20 cells. IL-2 produced was measured using the IL-2–dependent cell line HT-2 (39).

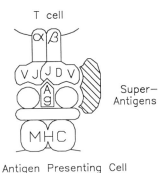

Fig. 3.   Schematic of MHC presentation of superantigens.

These data lead us to conclude that both self- and foreign superantigens bind to a similar site on the V_β element on the side of the TCR. Figure 3 shows a schematic of how we envisage the trimolecular complex of TCR–superantigen–MHC.

REFERENCES

1. Kronenberg, M., Siu, G., and Hood, L. E. (1986) *Annu. Rev. Immunol.* **4,** 529–591.
2. Bevan, M., and Fink, P. (1978) *Immunol. Rev.* **42,** 3–19.
3. Zinkernagel, R., Callahan, G., Althage, A., Cooper, S., Klein, P., and Klein, J. (1978) *J. Exp. Med.* **147,** 882–896.
4. Marrack, P., and Kappler, J. W. (1988) *Immunol. Today* **9,** 308–315.
5. Blackman, M. A., Marrack, P., and Kappler, J. (1989) *Science* **244,** 214–217.
6. Benoist, C., and Mathis, D. (1989) *Cell* **58,** 1027–1033.
7. Bill, J., and Palmer, E. (1989) *Nature* **341,** 649–651.
8. Festenstein, H. (1973) *Transplant. Rev.* **15,** 62–88.
9. Festenstein, H. (1974) *Transplantation* **18,** 555–557.
10. Janeway, C., Lerner, E., Jason, J., and Jones, B. (1980) *Immunogenetics* **10,** 481–497.
11. DeBreuil, P. C., Caillol, D. H., and Lemonnier, F. A. (1982) *J. Immunogenet.* **9,** 11–22.
12. Kappler, J. W., Staerz, U., White, J., and Marrack, P. (1988) *Nature* **332,** 35–40.
13. MacDonald, H. R., Schneider, R., Lees, R. K., Howe, R. C., Acha-Orbea, H., Festenstein, H., Zinkernagel, R. M., and Hengartner, H. (1988) *Nature* **332,** 40–45.
14. Pullen, A. M., Marrack, P., and Kappler, J. W. (1988) *Nature* **335,** 796–801.
15. Fink, P., Matis, L., McElligott, D., Bookman, M., and Hedrick, S. (1986) *Nature* **321,** 219–226.
16. Winoto, A., Urban, J., Lan, N., Goverman, J., Hood, L., and Hansburg, D. (1986) *Nature* **324,** 679–682.
17. Kappler, J., Wade, T., White, J., Kushnir, E., Blackman, M., Bill, J., Roehm, R., and Marrack, P. (1987) *Cell* **49,** 263–271.

18. Marrack, P., and Kappler, J. W. (1988) *Nature* **332,** 840–842.
19. Bill, J., Appel, V., and Palmer, E. (1988) *Proc. Natl. Acad. Sci. USA* **85,** 9184–9188.
29. Bill, J., Kanagawa, O., Woodland, D., and Palmer, E. (1989) *J. Exp. Med.* **169,** 1405–1419.
21. Woodland, D., Happ, M. P., Bill, J., and Palmer, E. (1989) *Science* **247,** 964–967.
22. Tomonari, K., and Lovering, E. (1988) *Immunogenetics* **28,** 445–451.
23. Happ, M. P., Woodland, D., and Palmer, E. (1989) *Proc. Natl. Acad. Sci. USA* **86,** 6293–6296.
24. Okada, C., and Weissman, I. L. (1989) *J. Exp. Med.* **169,** 1703–1719.
25. Pullen, A., Marrack, P., and Kappler, J. (1989) *J. Immunol.* **142,** 3033–3037.
26. Kappler, J. W., Roehm, N., and Marrack, P. (1987) *Cell* **49,** 273–280.
27. Marrack, P., and Kappler, J. W. (1990) *Science* **248,** 705–711.
28. Janeway, C. A., Jr., Yagi, J., Conrad, P., Katz, M., Vroegop, S., and Buxser, S. (1989) *Immunol. Rev.* **107,** 61–88.
29. White, J., Herman, A., Pullen, A. M., Kubo, R., Kappler, J. W., and Marrack, P. (1989) *Cell* **56,** 27–35.
30. Callahan, J., Herman, A., Kappler, J. W., and Marrack, P. (1990) *J. Immunol.* **144,** 2473–2479.
31. Kappler, J. W., Kotzin, B., Herron, L., Gelfand, E., Bigler, R., Boylston, A., Carrel, S., Posnett, D., Choi, Y., and Marrack, P. (1989) *Science* **244,** 811–813.
32. Choi, Y., Kotzin, B., Herron, L., Callahan, J., Marrack, P., and Kappler, J. W. (1989) *Proc. Natl. Acad. Sci. USA* **86,** 8941–8945.
33. Pullen, A. M., Potts, W., Wakeland, E. K., Kappler, J. W., and Marrack, P. (1990) *J. Exp. Med.* **171,** 49—62.
34. Wade, T., Bill, J., Marrack, P. C., Palmer, E., and Kappler, J. W. (1988) *J. Immunol.* **141,** 2165–2167.
35. Marrack, P., Blackman, M., Kushnir, E., and Kappler, J. W. (1990) *J. Exp. Med.* **171,** 455–464.
36. Pullen, A. M., Wade, T., Marrack, P., and Kappler, J. W. (1990) *Cell* **61,** 1365–1374.
37. Chothia, C., Boswell, D. R., and Lesk, A. M. (1988) *EMBO J.* **7,** 3745–3755.
38. Choi, Y., Herman, A., DiGiusto, D., Wade, T., Marrack, P., and Kappler, J. W. (1990) *Nature* **346,** 471–473.
39. Kappler, J., Skidmore, B., White, J., and Marrack, P. (1981) *J. Exp. Med.* **153,** 1198–1214.

13

Sequential Occurrence of Positive and Negative Selection during T Lymphocyte Maturation

CYNTHIA J. GUIDOS AND IRVING L. WEISSMAN
Department of Pathology,
Stanford University Medical School
Stanford, California 94305

INTRODUCTION

Genes controlling the immune functions of mature T lymphocytes are regulated by signaling through clonally diverse T cell antigen receptors (TCRs) on binding cell surface complexes of foreign antigenic peptides and major histocompatibility complex (MHC) proteins (1). In addition, TCR-mediated signals specify cell fates during T cell maturation in the thymus (2). Precursors bearing αβ TCR have three possible developmental fates: some mature into TCR^{hi} cells that express either the CD4 or CD8 MHC coreceptor molecules, but the majority fail to complete the maturation process and die *in situ* (3). Many studies have shown that TCR specificity is a critical determinant of thymocyte cell fate. Precursors bearing TCRs that can bind MHC molecules expressed on thymic stromal cells receive positive signals that result in rescue from programmed cell death and commitment to either the CD4 or CD8 lineage. However, precursors bearing TCRs that could potentially mediate effective T cell responses to self-peptide/self-MHC complexes are clonally deleted by a process called negative selection (reviewed in refs. 2 and 4). To survive this intrathymic selection process, precursors apparently must bear TCRs specific for nonself peptides bound to self-MHC molecules, resulting in a mature T cell repertoire that is maximally diverse but minimally autoreactive.

Molecular Mechanisms of
Immunological Self-Recognition

In order to study the molecular mechanisms by which TCR-mediated signal transduction activates distinct genetic pathways in immature and mature T cells, TCR$^+$ precursors that are the targets of positive and negative selection must be identified. The CD4$^+$ and CD8$^+$ T cell lineages both develop in the thymus from TCR$^-$ CD4-8$^-$ progenitors, via TCR^{lo} CD4$^+$8$^+$ blast cell intermediates (5). Several studies examining the frequency of self-reactive TCR V$_\beta$ in normal and $\alpha\beta$ TCR transgenic mice have implicated CD4$^+$8$^+$ thymocytes as targets of negative selection (6–8), but the findings have not been universal (9–12). Moreover, T cell maturation in TCR transgenic mice is abnormal in many respects, raising uncertainties about the conclusions derived. Finally, none of the studies reported to date have considered the heterogeneity of CD4$^+$8$^+$ thymocytes: only the blast subset (about 10% of the total) contains precursors for mature T cells (5). Most CD4$^+$8$^+$ thymocytes are postmitotic (nonmature) cells that die without further maturation. In fact, no other thymocyte subsets develop from purified small CD4$^+$8$^+$ thymocytes after intrathymic transfer to Thy-1 congenic hosts, and the cells die over a 2–4-day period, whereas both mature (CD4$^+$8$^-$ and CD4-8$^+$) and nonmature (small CD4$^+$8$^+$) thymocytes develop from intrathymically transferred CD4$^+$8$^+$ blasts (5). In this study, we have assessed the influence of TCR specificity on the developmental fate of CD4$^+$8$^+$ thymic blasts. Although TCR specificity has traditionally been determined by functional assays, recent studies using monoclonal antibodies and DNA hybridization probes specific for particular V$_\beta$ segments have revealed striking correlations between V$_\beta$ usage and TCR specificity. For example, most CD4$^+$ T cells expressing V$_\beta$6, V$_\beta$7, or V$_\beta$8.1 preferentially recognize determinants encoded by the minor lymphocyte stimulating (Mls-1^a) locus, and those expressing V$_\beta$5, V$_\beta$11, or V$_\beta$17 preferentially recognize MHC I-E molecules, (10, 13–16). Therefore, we used V$_\beta$-specific monoclonal antibodies to provide a phenotypic means of identifying the TCR specificity of immature TCR^{lo} CD4$^+$8$^+$ blasts and to follow their maturation into CD4$^+$ and CD8$^+$ T cells in the presence and absence of the relevant TCR ligand. The results indicate that positive selection and negative selection of precursors specific for these antigens occur sequentially during T cell maturation and that the cellular targets of each event are distinguishable by differences in surface levels of CD4, CD8, and TCR.

RESULTS AND DISCUSSION

The maturational phase during which thymocytes first express surface TCR, becoming potential targets of repertoire selection, has not been unequivocally identified. Studies in the rat indicate that the outer cortical CD4-8$^+$ precursors of CD4$^+$8$^+$ thymocytes are TCR^{lo} (17), but the analogous murine subset has been reported to be TCR$^-$ (18). In contrast, phenotypic observations in mice and humans have shown that only about 50% of total CD4$^+$8$^+$ thymocytes are discernibly TCR^{lo}, suggesting that thymocytes acquire surface TCRs sometime after they become CD4$^+$8$^+$ (19, 20). However, we have used a sensitive staining technique to show that the majority of precursor (blast) and small CD4$^+$8$^+$ thymocytes express 10- to 20-fold fewer surface TCR molecules than mature T cells (12). To probe the functional status of TCR on CD4$^+$8$^+$ blasts, we measured changes in intracellular Ca^{2+} concentration in response to cross-linking by an anti-CD3 monoclonal antibody. This response is one of the earliest measurable indices of TCR-mediated signal transduction and has been well studied in mature T cells (21) and unfractionated CD4$^+$8$^+$ thymocytes (19, 20), but not in the precursor blast subset. Thymocytes were loaded with the Ca^{2+}-sensitive dye Indo-1 prior to staining with anti-CD4 (phycoerythrin), anti-CD8 (fluorescein isothiocyanate, (FITC), and anti-CD3. Multiparameter flow cytometry was used as described (22) to measure Indo-1 fluorescence in various thymocyte subsets before and after cross-linking of surface CD3. As expected, CD4$^+$8$^-$ thymocytes displayed a large net increase in intracellular free [Ca^{2+}] (approximately 6-fold) after CD3 ligation (Fig. 1). CD4$^+$8$^+$ blasts also responded to CD3 cross-linking, and the net [Ca^{2+}] increase was only 3-fold lower than in CD4$^+$8$^-$ thymocytes, despite expression of 10- to 20-fold fewer surface CD3 molecules. In contrast, small CD4$^+$8$^+$ thymocytes were heterogeneous in their responsiveness. In about half the cells (Fig. 1, population a), the response was similar in magnitude and kinetics to that of CD4$^+$8$^+$ blasts, whereas the remainder of the cells (Fig. 1, population b) showed no significant response. The developmental significance of this observation is unclear, because these cells are irreversibly committed to intrathymic death and cannot be targets of positive selection (12). It is possible that TCR-mediated signal transduction has developmental consequences only for the immature TCR^{lo} CD4$^-$8$^+$ and CD4$^+$8$^+$ blast subsets. Some nonmature small CD4$^+$8$^+$ cells may transiently retain residual responsiveness to TCR-mediated signals, whereas those in the later phases of programmed cell death may become unresponsive. Thus, the majority of TCR^{lo} CD4$^+$8$^+$ blasts possess functional surface TCR, but this approach cannot elucidate the functional consequences (positive or negative selection) of TCR-mediated signaling at this maturational stage. Correlations between TCR-mediated Ca^{2+} increases and susceptibility to anti-TCR–mediated deletion of thymocytes in fetal thymus organ culture (23) do not resolve this issue, because potentially responsive cells in these cultures include TCR^{lo} CD4$^-$8$^+$ precursors as well as immature (blast) and nonmature (small) TCR^{lo} CD4$^+$8$^+$ thymocytes.

Many studies have reported that negative selection occurs at the CD4$^+$8$^+$ maturational stage, based on the lower frequency of particular TCR-V$_\beta$ segments among total CD4$^+$8$^+$ thymocytes in deleting strains (e.g., Mls-1^a,I-E$^+$)

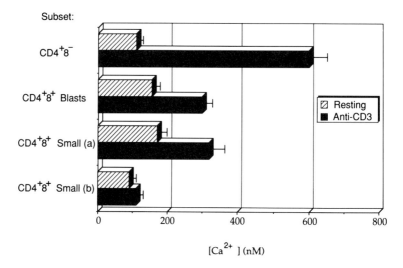

Fig. 1. TCR-mediated changes in intracellular [Ca²⁺] in CD4⁺8⁺ and CD4⁺8⁻ thymocyte subsets. The sources and fluorochrome modifications of monoclonal antibodies used in this study have been described (5, 26). For Ca²⁺ flux measurements, AKR/J neonatal (1-week-old) thymocytes (10^7/ml) were incubated in Hanks' balanced salt solution containing 10 mM HEPES, 2% calf serum, and 3 µM Indo-1 (Molecular Probes, Eugene, OR) for 45 min at 37°C. Cells were then washed twice, incubated for 45 min at 4°C with phycoerythrin-GK1.5 (anti-CD4, Becton Dickinson, Mountain View, CA), FITC-53-6.7 (anti-CD8), and 500A2 (anti-CD3) culture supernatant and stored on ice until the time of analysis. Immediately before fluorescence measurements were carried out on a dual argon laser (488-nm and 340-nm excitation) FACStar Plus (Becton Dickinson), cells were warmed to 37°C for 3 min. Indo-1 emissions (collected at 540 nm and 405 nm) were measured for 3 min before (resting) and for 10 min after cross-linking of surface CD3 with antihamster IgG. 540/405 fluorescence ratios were converted to [Ca²⁺] (nM) using a calibration curve generated as described (22). Maximal anti-CD3 stimulated responses are shown. Fluorescence signals were computer gated to determine the intracellular [Ca²⁺] in the indicated subsets. Addition of antihamster IgG to thymocytes not preincubated with 500A2 did not raise intracellular [Ca²⁺] above resting levels. Approximately 50% of small CD4⁺8⁺ thymocytes showed a TCR-induced increase in intracellular [Ca²⁺] (population a), whereas the remainder (population b) were unresponsive.

relative to nondeleting strains (Mls-1ᵇ, I-E⁻) (reviewed in ref. 2). However, there have been conflicting reports, and we found no phenotypic evidence for strain-specific differences in V$_β$3, V$_β$5, V$_β$6, V$_β$7, V$_β$11, or V$_β$17 frequency among either small or blast CD4⁺8⁺ thymocytes (refs. 12, 24, and data not shown). Furthermore, we showed that intrathymic transfer of CD4⁺8⁺ blasts from a deleting (Mls-1ᵃ) to a nondeleting (Mls-1ᵇ) host could rescue the production of V$_β$6⁺ (TCRʰⁱ) mature thymocytes (12), suggesting that no irreversible negative signals have been delivered to CD4⁺8⁺ blasts. Clonal deletion must therefore occur during the final transition from TCR^{lo} CD4$^+$8$^+$ blasts to TCR^{hi} CD4$^+$ or CD8$^+$ T cells. To identify thymocytes corresponding to this late maturational stage, we used three- and four-color flow cytometry to characterize CD4, CD8, and TCR/CD3 expression as CD4$^+$ and CD8$^+$ cells developed from CD4$^+$8$^+$ blasts after intrathymic transfer into Thy-1 congenic hosts. Three to four days after transfer, progeny displaying CD4^{hi}8^{lo} or CD4^{lo}8^{hi} "transitional" phenotypes were observed (Fig. 2). Thymocytes expressing transitional levels of CD4 or CD8 can also be identified among neonatal or young adult thymocytes (12). Interestingly, such lineage-committed intermediates also express transitional levels of TCR and CD5, another T cell differentiation antigen (Fig. 3). Many CD4^{hi} cells bearing transitional levels of CD8 (CD8^{med} or CD8^{lo}) express 5- to 7-fold higher levels of TCR and CD5 than TCR^{lo} CD4$^+$8$^+$ blasts. Such cells are referred to as TCR^{med}. The final transition from CD4^{hi}8^{lo} to CD4^{hi}8$^-$ is accompanied by a further 2- to 3-fold increase in surface TCR and CD5 density (TCR^{med} to TCR^{hi}). Thymocytes committed to the CD8 lineage can be similarly identified by their CD4$^{med/lo}$ CD8^{hi} and TCR/CD5^{med} phenotype (Fig. 3, bottom). It is interesting that the final TCR^{med}-to-TCR^{hi} transition of the CD8 lineage involves a greater increase in surface TCR density than the

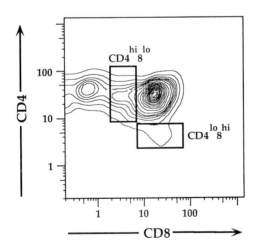

Fig. 2. CD4$^+$8$^+$ blast progeny display transitional phenotypes. CD4$^+$8$^+$ blasts from C57B16/Ka (Thy-1.2) mice were purified by unit gravity sedimentation followed by cell sorting and injected intrathymically into unirradiated C57B16/Ka (Thy-1.1) host mice as previously described (12). Three and one-half days later, Thy-1.2$^+$ donor-derived cells were recovered using immunomagnetic beads and analyzed for CD4 vs. CD8 expression on a highly modified dual (argon and rhodamine dye) laser FACS IV (Becton-Dickinson) with four-decade logarithmic amplifiers. Excitation wavelengths were 488 nm (argon laser) and 598 nm (dye laser). CD4^{lo} and CD8^{lo} designations were made by comparison with lymph node CD4/8 profiles as described (12).

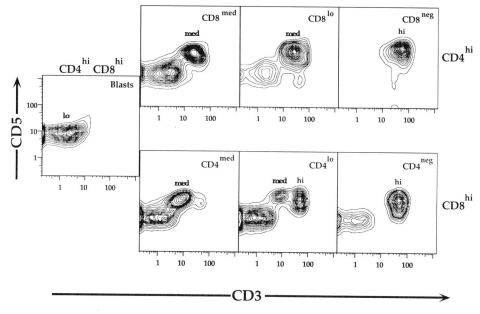

Fig. 3.   CD3 vs. CD5 expression during late stages of T cell maturation. Thymocytes were stained with anti-CD4 (phycoerythrin-GK1.5), anti-CD8 (biotin-53-6.7/avidin-allophycocyanin), anti-CD3 (500A2 culture supernatant/antihamster–Texas Red), and anti-CD5 (FITC-53-7.3). Fluorescence data were collected as described for Fig. 2, except that the wavelength of the dye laser was raised to 605 nm. Data were analyzed using the FACS/DESK software program as described (12) to generate the CD3 vs. CD5 profiles (5% probability plot) on each subset shown. Transitional (medium and low) CD4 and CD8 designations were made as described (12).

analogous step in the CD4 lineage, but the developmental and functional significance of this is unknown.

Figure 4 shows a complementary analysis of the CD4/8 phenotypes of thymocytes with low (*top panel*), medium or transitional (*middle panel*), and high (*bottom panel*) levels of surface TCR. The frequency of CD4^{hi} thymocytes bearing transitional (medium and low) levels of CD8 is more than 10-fold higher among TCR^{med} thymocytes than among the TCR^{lo} subset. The final TCR^{med}-to-TCR^{hi} transition is marked by a 10-fold increase in mature CD4^{+}8^{-} cells and a 2- to 3-fold relative decrease in CD4^{+}8^{med} cells. Thus, thymocytes bearing transitional levels of TCR are enriched in cells bearing transitional levels of CD8, and vice versa. These observations suggest that down-regulation of CD4 or CD8 and up-regulation of TCR and CD5 occur simultaneously during the terminal phases of T cell maturation.

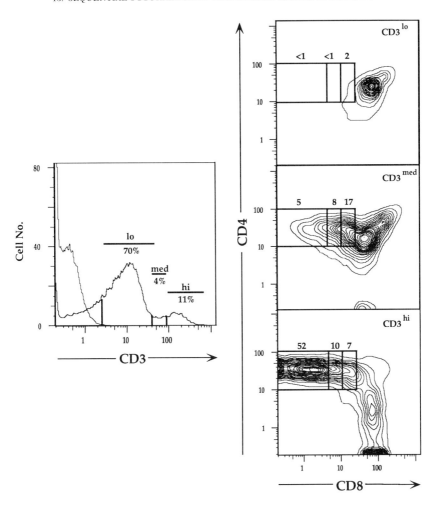

Fig. 4. CD4/8 phenotypes of cells during the TCR^{lo}-to-TCR^{hi} transition. AKR thymocytes were stained as described for Fig. 1. Thymocytes were designated CD3^{hi} by comparison to the level of CD3-specific fluorescence of lymph node cells. The CD4/8 profiles were generated by computer gating of the CD3 fluorescence signals. The three boxed areas of each CD4/8 profile show gates used to define (from left to right) CD4$^+$8$^-$, CD4$^+$8^{lo}, and CD4$^+$8^{med} populations within each CD3 subset. CD8 designations were made as described (12).

Lineage commitment and maturation of TCR^{lo} CD4$^+$8$^+$ blasts to the TCR^{med} stage most likely occur as a result of positive selection, since V$_\beta$17 frequency at this maturational stage is directly correlated with the efficiency of various MHC haplotypes to positively select this receptor (12). Moreover, the outcome of negative selection is not yet apparent among TCR^{med} transitional thymocytes, since V$_\beta$6 and V$_\beta$17 frequencies are often high in deleting (Mls-1^a and I-E$^+$) relative to nondeleting strains (12). Figure 5 shows a similar analysis of V$_\beta$11 frequency among immature TCR^{lo} CD4$^+$8$^+$ blasts, TCR^{med} transitional thymocytes, and mature TCR^{hi} CD4$^+$ and CD8$^+$ thymocytes in the presence and absence of the I-E deleting element. The frequency of V$_\beta$11 among TCR^{med} thymocytes of both lineages was as high in B10.BR (I-E$^+$) mice as in BA and B10.A(4R) (I-E$^-$) mice, yet deletion in both the CD4 and CD8 lineages was readily apparent among TCR^{hi} thymocytes in B10.BR mice (Fig. 5). These observations are most consistent with the idea that negative selection of autoreactive thymocytes occurs only after positive selection results in lineage commitment and acquisition of a TCR^{med} transitional phenotype. This hypothesis is supported by our analysis of V$_\beta$17 frequency among lineage-committed TCR^{med} thymocytes in F1 strains that coexpress the I-A^q (SWR) positive selecting element but that differ in expression of the I-E encoded deleting element. As shown in Fig. 6, I-E expression has no influence on V$_\beta$17 frequency among

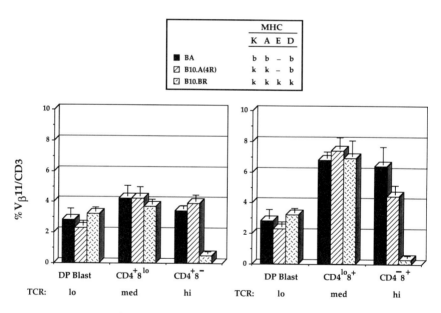

Fig. 5. Frequency of V$_\beta$11 among TCR^{lo} CD4$^+$8$^+$ blasts, TCR^{med} transitional cells, and TCR^{hi} mature thymocytes. Young adult thymocytes of the indicated mouse strains were analyzed as described for Fig. 3, except that anti-V$_\beta$11 (RR3-15 culture supernatant/FITC–antirat Ig) was used instead of anti-CD5. Percentages refer to V$_\beta$11$^+$ (low, medium, or high) relative to total CD3$^+$ (low, medium, or high) in each CD4/8 subset. CD4$^+$8^{lo} thymocytes include CD8^{med} and CD8^{lo} cells as they are defined in Fig. 3, and CD4^{lo}8$^+$ cells include CD4^{med} and CD4^{lo} thymocytes.

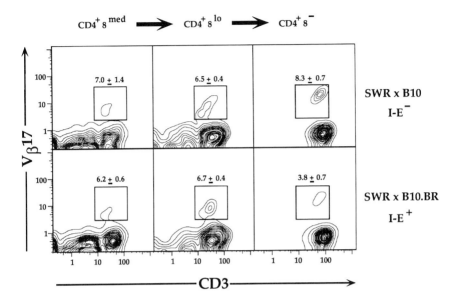

Fig. 6. $V_\beta 17$ frequency among transitional and mature CD4$^+$ thymocytes in I-E$^-$ and I-E$^+$ H-2^q haplotypes. Young adult thymocytes from each F$_1$ strain were stained as described for Fig. 3, except that FITC-KJ23 (anti-$V_\beta 17$) was substituted for anti-CD5. The profiles show $V_\beta 17$ vs. CD3 expression for the indicated maturational stages from one mouse of each haplotype. The numbers above each box refer to the percentage (mean ± SD, 3 animals per haplotype) of $V_\beta 17^{med}$ (*left and middle panels*) or $V_\beta 17^{hi}$ cells (*right panels*) relative to total CD3^{med} or CD3^{hi} cells in each subset.

TCR^{med} CD4$^+$8^{med} or CD4$^+$8^{lo} thymocytes. However, $V_\beta 17$ frequency among TCR^{hi} CD4$^+$8$^-$ thymocytes was 50% lower in I-E$^+$ mice. We conclude from these studies that negative selection occurs during the final TCR^{med}-to-TCR^{hi} transition in both the CD4 and CD8 lineages.

These observations have two implications. First, the data suggest that positive selection and negative selection occur sequentially during T cell maturation. It could be argued that deleting autoreactive cells from only those few that have been positively selected based on self-MHC restriction specificity is the most efficient way to achieve a maximally diverse and minimally autoreactive T cell repertoire. Second, the data suggest that susceptibility to TCR-mediated negative selection may be contingent on prior positive selection. According to this view, the response of TCR$^+$ immature thymocytes to TCR-mediated signals would be developmentally programmed such that only one outcome is possible during a particular maturational phase. Thus, signaling through the TCR would result in positive selection for TCR^{lo} CD4$^+$8$^+$ blasts, negative selection for TCR^{med} CD4$^+$8^{lo} or CD4^{lo}8$^+$ transitional cells, and proliferation and effector functions for TCR[hi] mature cells. Because the vast majority of dying thymocytes are postmitotic TCR[lo] CD4+8+ cells, this model implies that the major cause of cell death is failure of positive selection (which would rescue cells from programmed cell death), rather than negative selection. Others have attempted to explain the selective localization and opposite outcomes of positive and negative selection by postulating that cortical and medullary thymic stromal cells provide different non-TCR signals (4), or qualitatively different self-antigen/MHC complexes (11). However, this could merely reflect the fact that, under normal circumstances, medium and high TCR/CD3 levels are attained only after cortical precursors have migrated to the juxtamedullary region. Therefore, despite the presence of at least some extrathymic self-antigens on thymic cortical epithelial cells (25), TCR[lo] precursor cells would not be clonally deleted under normal circumstances. However, cortical deletion of CD4+8+ thymocytes apparently occurs following temporally (and presumably spatially) premature high-level surface TCR expression in the thymic cortex of mice expressing self-reactive αβ TCR transgenes.

In summary, our results suggest that maturation of TCR[lo] CD4+8+ blast cells into TCR[hi] CD4+ or CD8+ thymocytes occurs via TCR[med] intermediates, and that the consequences of TCR-mediated signal transduction differ at each maturational phase. We propose that TCR density is likely to be a critical determinant of this differential responsiveness, but because the magnitude of the TCR-stimulated Ca^{2+} response in immature CD4+8+ blasts was only 2- to 3-fold lower than that in mature thymocytes, despite a 10- to 20-fold difference in surface TCR density, the influence of TCR-mediated signals on developmental choices may not be strictly quantitative. In addition, response thresholds may vary according to maturational stage to ensure that not all positively selected cells are clonally deleted. The data support a view of TCR-mediated selective events during intrathymic maturation that differs substantially from the one suggested by studies of αβ TCR transgenic mice. Although developmentally deregulated expression of TCR transgenes has yielded interesting and informative results that bear on TCR-mediated alteration of normal thymocyte fates, the usefulness of these systems as paradigms of normal intrathymic maturation may be limited.

REFERENCES

1. Ullman, K. S., Northrop, J. P., Verweij, C. L., and Crabtree, G. R. (1990) *Annu. Rev. Immunol.* **8,** 421–452.
2. Schwartz, R. H. (1989) *Cell* **57,** 1073–1081.
3. Adkins, B., Mueller, C., Okada, C., Reichert, R., Weissman, I. L., and Spangrude, G. J. (1987) *Annu. Rev. Immunol.* **5,** 325–365.
4. Sprent, J., Lo, D., Gao, E., and Ron, Y. (1988) *Immunol. Rev.* **101,** 173–190.

5. Guidos, C. J., Weissman, I. L., and Adkins, B. (1989) *Proc. Natl. Acad. Sci. USA* **86,** 7542–7546.
6. Fowlkes, B. J., Schwartz, R. H., and Pardoll, D. M. (1988) *Nature* **334,** 620–623.
7. MacDonald, H. R., Hengartner, H., and Pedrazzini, T. (1988) *Nature* **335,** 174–176.
8. White, J. Herman, A., Pullen, A. M., Kubo, R., Kappler, J. W., and Marrack, P. (1989) *Cell* **56,** 27–35.
9. Kappler, J. W., Roehm, N., and Marrack, P. (1987) *Cell* **49,** 273–280.
10. Kappler, J. W., Staerz, U., White, J., and Marrack, P. C. (1988) *Nature* **332,** 35–40.
11. Marrack, P., Lo, D., Brinster, R., Palmiter, R., Burkly, L., Flavell, R. H., and Kappler, J. (1988) *Cell* **53,** 627–634.
12. Guidos, C. J., Danska, J. S., Fathman, C. G., and Weissman, I. L. (1990) *J. Exp. Med.* **172,** 835–845.
13. Kappler, J. W., Wade, T., White, J., Kushnir, E., Blackman, M., Bill, J., Roehm, N., and Marrack, P. (1987) *Cell* **49,** 263–271.
14. MacDonald, H. R., Schneider, R., Lees, R. K., Howe, R. C., Acha-Orbea, H., Festenstein, H., Zinkernagel, R. M., and Hengartner, H. (1988) *Nature* **322,** 40–45.
15. Bill, J., Kanagawa, O., Woodland, D. L., and Palmer, E. (1989) *J. Exp. Med.* **169,** 1405–1419.
16. Okada, C. Y., and Weissman, I. L. (1989) *J. Exp. Med.* **169,** 1703–1719.
17. Hunig, T. (1988) *Eur. J. Immunol.* **18,** 2089–2092.
18. MacDonald, H. R., Budd, R. C., and Howe, R. C. (1988) *Eur. J. Immunol.* **18,** 519–523.
19. Havran, W. L., Poenie, M., Kimura, J., Tsien, R., Weiss, A., and Allison, J. P. (1987) *Nature* **330,** 170–173.
20. Weiss, A., Dazin, P. F., Sheilds, R., Man Fu, S., and Lanier, L. L. (1987) *J. Immunol.* **139,** 3245—3250.
21. Imboden, J. B. and Weiss, A. (1988) *Prog. Allergy* **42,** 246–279.
22. Parks, D. R., Nozaki, T., Dunne, J. F., and Peterson, L. L. (1987) *Cytometry* **8** (Suppl. 1), 104.
23. Helman Finkel, T., Cambier, J. C., Kubo, R. T., Born, W. K., Marrack, P., and Kappler, J. W. (1989) *Cell* **58,** 1047–1054.
24. Okada, C. Y., Holzmann, B., Guidos, C., Palmer, E., and Weissman, I. L. (1990) *J. Immunol.* **144,** 3473–3477.
25. Lorenz, R. G. and Allen, P. M. (1989) *Nature* **337,** 560–562.
26. Guidos, C. J., Weissman, I. L., and Adkins, B. (1989) *J. Immunol.* **142,** 3773–3780.

14

Tolerance Induction in the Peripheral Immune System

MATT WISE, RICHARD BENJAMIN, SHIXIN QIN,
STEPHEN COBBOLD, AND HERMAN WALDMANN
Immunology Division,
Department of Pathology,
Cambridge University,
Cambridge CB2 1QP, England

INTRODUCTION

Throughout life it is essential that the immune system remains alert to antigens from the outside world, while avoiding self-reactivity (1). Clonal anergy, deletion, and suppressor cells have all been evoked as the mechanism by which antiself lymphocytes are controlled (2). In large part the problem becomes one of how T lymphocytes are controlled, because without these no harmful self-reactivity is possible. Although much of T cell self-tolerance arises centrally (in the thymus) through deletion of T cells with receptors for self-peptides, the T cells that leave the thymus still require a mechanism(s) to ensure tolerance to tissue-restricted peptides. In part, peripheral tolerance could be explained by clonal anergy, although this begs the question of how the T cell opted for anergy rather than immunity. The peripheral tolerance mechanisms, although minor in their contribution to self-tolerance, remain the ones which are most relevant to clinical exploitation. After all, the thymus is not easily accessible therapeutically.

We have sought, therefore, to follow a model of peripheral tolerance induced within an adult animal, with a view to understanding mechanism. We have previously shown (3) that T cell tolerance can be induced in mice to the foreign protein human gamma globulin (HGG) by giving that protein under the umbrella of CD4 mAb therapy. By choosing a rat CD4 mAb that cannot kill T cells *in vivo*, and by using adult thymectomized mice, we can guarantee that any tolerance must be determined through the peripheral T cell system alone.

Molecular Mechanisms of
Immunological Self-Recognition

Using this simple system we show that peripheral tolerance is antigen specific and functionally permanent. It arises and is maintained solely within the CD4+ T cell subpopulation, as CD8-depleted mice are equivalently tolerizable and maintain their state of tolerance indefinitely. Perhaps the most remarkable feature of the model is that within tolerant animals the CD4 compartment contains cells capable of "resisting" normal (but not primed) T cells from breaking that tolerant state, when these cells are transferred from a normal donor to the tolerant host. This suggests that the tolerant CD4+ T cells are not inert or nonfunctional but are capable of interacting to inhibit the function of "nontolerant" cells.

Furthermore, we demonstrate that it is possible to induce peripheral tolerance in mice previously sensitized to HGG.

MATERIALS AND METHODS

CBA/Ca mice were used throughout, and all CD4 and CD8 antibodies have been previously described (4, 5). All serum antibody responses were measured by enzyme-linked immunosorbent assay (ELISA) as described previously (3, 5). Starting dilutions of test sera were 1:20 (i.e., a $\log_2$ titer of 1=1:20 serum dilution). Tolerance to HGG was induced using an essentially similar protocol to one previously described (3, 5).

RESULTS

Induction and Maintenance of Peripheral Tolerance to HGG

We have previously shown that the CD4 mAb YTS 177 does not deplete target cells, although it can modulate the level of CD4 expressed on the cell surface (5, 6). By using this mAb to induce tolerance to HGG we can maintain normal numbers of peripheral T cells and so examine the role, if any, of the thymus in tolerance induction and maintenance. Tolerized adult thymectomized mice challenged at 6 or 12 weeks with HGG were unable to respond, demonstrating that the thymus was not essential for tolerance induction (5). Control euthymic and thymectomized mice that received mAb only and no tolerizing dose of HGG were, in contrast, able to respond. This then rules out nonspecific immunosuppression.

In the past we have noted that when the challenge dose of HGG was delayed for many months, euthymic mice would lose tolerance, while giving the chal-

TABLE I
Permanent Peripheral Tolerance in the Absence of a Thymus[a]

Group	Thymus	Tolerogen (HGG)	Antibody titer ($\log_2$) to HGG (SD) challenge on day 213
1	Yes	No	8.46 (2.2)
2	No	No	5.72 (2.0)
3	Yes	Yes	6.98 (0.84)
4	No	Yes	1.57 (0.82)

[a]Groups of 5–7 mice were used. Titers represent the geometric mean of groups immunized with HGG (0.5 mg of aggregated material) at day 213 and boosted with the same dose a week later. Adult thymectomized or euthymic mice were given 2 mg of CD4 antibody [YTS 177 on days –1 (i.v.), 0 and +1 (i.p.)]. Some mice were given 1 mg of HGG on day 0. Controls received no tolerogen (annotated from ref. 5).

lenge earlier "reinforced" that tolerant state. In contrast, thymectomized mice remain permanently unresponsive (>200 days) (Table I).

We conclude that in euthymic animals tolerance is broken by new T cells that exit the thymus into an antigen-free periphery.

CD8 T Cells Are Not Involved in the Induction or Expression of HGG Tolerance

The fact that euthymic mice tolerized to HGG were susceptible to reinforcement by further exposure to antigen (7) suggests that some antigen-specific regulatory mechanism might be operating. Classically, active regulation/suppression has been attributed to the CD8 subset. Previous experiments have shown that tolerance to HGG induced by depleting antibodies was not dependent on CD8[+] T cells (7). We were able to confirm the same for peripheral tolerance induced with the nondepleting CD4 antibody. Adult thymectomized mice were given the tolerizing CD4 protocol (days 0, 1, 2) as before and depleted of CD8 T cells (days 4, 5) just before the tolerogen was administered on day 8. A further group tolerized at the start of the experiment had their CD8 T cells depleted by antibodies given on days 41 and 43 (just before rechallenge with HGG). Seven of eight mice from the "early" and all eight mice from the "late" CD8 depletion groups were tolerant (Table II).

This rules out any role of the CD8 subset in the induction, maintenance, or readout phases of tolerance.

TABLE II
CD8 T Cells Not Necessary for Peripheral Tolerance Induced with CD4 mAb[a]

Group	CD4 mAb	CD8 depletion	Tolerogen	Anti-HGG titer $(\log_2) \pm$ SD day 61
1	Yes	No	No	7.8 (1.7)
2	Yes	Early	Yes	1.5 (1.3)
3	Yes	Late	Yes	<1
4	Yes	No	Yes	<1
5	Yes	No	No	8.35 (1.8)

[a]All mice (8 per group) were challenged with 0.5 mg of aggregated HGG on days 43 and 54. CD8 depletion was with mAb YTS 169 at 2-mg doses (i.v./i.p.) on either days 4 and 5 (early) or days 41 and 43 (late). YTS 177 was given at 2 mg of protein (ammonium sulfate cut/ascites) on days 0 (i.v.), 1, and 2 (i.p.). HGG was given (heat-aggregated) at 1 mg (i.p.) on day 8.

Tolerant CD4⁺ T Cells Resist Breakdown of Tolerance by Normal T Cells

Although CD8 T cells are not necessary for generating or maintaining tolerance to HGG, the reinforcement of tolerance by repeated antigen (HGG) administration (7, 5) does suggest that some regulatory event is going on. In the past a common experimental approach to searching for "suppression" was to transfer cells from tolerant mice together with normal cells into a neutral recipient and observe whether tolerant cells could suppress the normal ones. Having previously failed to document suppression by such "transfer" (7), we opted to transfer normal cells into tolerant adult thymectomized animals to see if we could break the tolerant state. Table III shows that thymectomized mice rendered tolerant to HGG with YTS 177 remained tolerant after transfer of 50 million naive spleen cells. The same spleen cells could, however, elicit immune responses in "tolerized" recipients who had their CD4 T cells ablated with the depleting CD4 mAb YTS 191. In contrast, HGG-primed spleen cells were able to break resistance in the tolerant hosts. As a control, all recipients received ovalbumin. All groups made equivalent anti-OVA responses (data not shown).

"Resistance" must therefore be mediated by CD4 T cells in the tolerant host.

Tolerance Induction in T Cells Primed to HGG

The ability of "depleting" CD4 mAb therapy to induce tolerance to soluble protein antigens is well documented (3, 8, 9). However, previous priming prevented tolerance with the same protocols, despite profound immunosuppres-

TABLE III
Resistance in HGG-Tolerant Animals Mediated by CD4 T Cells[a]

Group	Donor status	Host CD4 ablation	Log$_2$ anti-HGG titer (SD)
1	Naive	No	1.13 (0.2)
2	HGG-primed	No	6.23 (1)
3	Naive	Yes	5.76 (1.07)
4	HGG-primed	Yes	7.34 (0.6)

[a]All adult thymectomized mice (5–7 per group), rendered tolerant to HGG as usual, were bled on day 26 and confirmed as making no anti-HGG response (not shown). A single intravenous dose of YTS 191 (rat IgG2b) was given to half the mice on day 41 to deplete their CD4$^+$ T cells. On day 62 approximately 50 million freshly harvested, pooled spleen cells were transferred to each mouse by intravenous injection. Primed donor mice had been immunized with 0.5 mg HGG in Freund's complete adjuvant and boosted 87 days later with 500 µg of soluble protein (i.p.) 9 days prior to harvest. All recipient mice were challenged with 0.5 mg ovalbumen on days 62 and 70. All serum antibody responses were measured on day 82.

sion (unpublished data). As combined protocols of "depleting followed by nondepleting" CD4 mAb therapy could tolerize animals previously primed to minor transplantation antigens (6), we investigated whether similar protocols could be used to tolerize to HGG.

Combination therapy ("depletion followed by blockade") resulted in complete failure of the mice to mount an anti-HGG response. However, no anti-HGG response could be detected in the control group receiving antibody alone! This raised the question of whether the control group was merely immunosuppressed (and not tolerant) or whether the control group was tolerant to either the HGG persisting from the original priming or HGG given in the challenge doses from day 33. All four groups of mice were rechallenged with 0.5 mg of aggregated HGG (i.p.) on days 66, 77, 114, and 121 with antibody measurements on days 84 and 129. Even on these later bleeds both groups of mice still failed to respond despite the fact that the time limit for T cell regeneration and antibody clearance had well passed. Clearly, we had elicited tolerance to HGG, but we could not establish which antigen dose had done it—the priming dose or the first or subsequent challenge doses?

DISCUSSION

Monoclonal antibodies to murine CD4 have exhibited potent therapeutic effects *in vivo* (reviewed in ref. 10). Early work capitalized on the observation that antibodies of the rat IgG2b isotype could efficiently exploit mouse effector mechanisms, causing depletion (4, 3, 7, 9). However, it became apparent that nondepleting doses of F (ab′)$_2$ of the same antibodies could also cause tolerance (7, 11, 12). This was an exciting observation because it suggested that the immune system might be tolerizable without need to kill any T cells. The emergence of a potent nondepleting rIgG$_{2a}$ CD4 mAb (YTS 177) has enabled us to study "tolerance without depletion" in more detail and to be able to address mechanism (5). It is clear from the data of Table I that the capacity to become tolerant and to sustain that tolerance resides wholly within the extrathymic peripheral T cell system.

The failure of thymectomized mice to respond to antigen even as late as 233 days after the initial antigen exposure suggests that the unresponsive state is long-lasting. The eventual return of antigen responsiveness in tolerized euthymic mice must then be due to the new T cells emerging from the thymus. Clonal deletion seems insufficient to explain the prolonged tolerant state, because responsiveness returns well after T cells have regenerated. The observation that tolerance can be reinforced indefinitely (5, 7), would seem to favor a regulatory mechanism. However, any such mechanism could not be dependent on CD8$^+$ cells, as we have demonstrated tolerance in their absence (Table II). Moreover, by transferring naive donor cells into thymectomized tolerant mice depleted of CD4$^+$ cells it was possible to attribute a regulatory role to the CD4$^+$ subset (Table IV). The capacity of "tolerant" cells to "resist" tolerance breakdown by naive but not primed T cells can be interpreted in one of two ways. It may be that the mechanism for regulation for naive cells fails to operate on primed cells, or that there is simply a quantitative element to regulation and that primed cells contain too many immunocompetent HGG-specific T cells.

As resistance has now been documented in antibody-mediated tolerance to bone marrow as well as to skin grafts (8, 5), it seems to be a general phenomenon. Elucidation of its mechanism may provide the single most important explanation for CD4 antibody-mediated tolerance and consequently permit directed therapies to potentiate such mechanisms in control of autoimmunity and graft rejection.

In the management of autoimmunity it will be essential to generate tolerance or "reprogram" a previously sensitized immune system. It is clear from the present studies with HGG and studies in other transplantation models (6) that tolerance can be induced in primed animals. In some cases this has not required any T cell depletion, although in the case of HGG tolerance some debulking of CD4 T cells was required. Given these initial findings, we need to know if "resistance" also operates in tolerance within a primed immune system. If there were a common theme to all forms of antibody-medicated tolerance, this would greatly simplify clinically targeted research to harness tolerance as a therapeutic process.

TABLE IV
CD4 Therapy to Induce Tolerance in Primed T-Cells[a]

Group	HGG at days 7, 14	CD4 Rx	Log$_2$ anti-HGG titer (SD)		
			Day 50	Day 84	Day 129
1	None	None	9.6 (0.8)	11 (1.25)	10.1 (1)
2	Yes	None	7.8 (1.2)	8.8 (1)	9.9 (1)
3	None	Yes	< 1	< 1	< 1
4	Yes	Yes	< 1	< 1	< 1

[a]All mice (8 per group) were primed with 50 μg of heat-aggregated HGG 26 days prior to antibody therapy. Experimental mice received 3×2 mg doses of YTS 191 (depleting) on days −2 (i.v.), −1, and 0 (i.p.), followed by 2 mg of YTS 177 (nondepleting) on days 2, 5, 7, 9, 12, 14, and 16 (i.p.). Antigen in the form of 1 mg of aggregated HGG was given on days 0, 7, and 14 by the intraperitoneal route. Control mice received no treatment, HGG only, or antibody only. All groups were bled on day 21 to assess antibody titers and were later challenged with 0.5 mg aggregated HGG on days 33 and 43, followed by measurement of the anti-HGG response on day 50.

ACKNOWLEDGMENTS

This work was supported by the Medical Research Council, UK, and the Arthritis and Rheumatism Council. M.W. was a recipient of a Wellcome scholarship.

REFERENCES

1. Ehrlich, P., and Morgenroth, J. (1900) In *The Collected Papers of Paul Ehrlich* (ed. F. Himmelweit, M. Marquardt, and H. D. Dale), Pergamon, London, 205–212.
2. Nossal, G. (1983) *Annu. Rev. Immunol.* **1**, 33–62.
3. Benjamin, R. J., and Waldmann, H. (1986) *Nature* **320**, 449–451.
4. Cobbold, S. P., Jayasuriya, A., Nash, A., Prospero, T. D., and Waldmann, H. (1984) *Nature* **312**, 548.
5. Qin, S., Wise, M., Cobbold, S. P., Leung, L., Kong, Y.-C. M., Parnes, J. R., and Waldmann, H. (1990) *Eur. J. Immunol.* **20**, 2737–2746.
6. Cobbold, S. P., Martin, G., and Waldmann, H. (1990) *Eur. J. Immunol.* **20**, 2747–2756.
7. Benjamin, R. J., +Qin, S., Wise, M. P., Cobbold, S. P., and Waldmann, H. (1988) *Eur. J. Immunol.* **18**, 1079–1088.
8. Qin, S., Cobbold, S. P., Benjamin, R. J., and Waldmann, H. (1989) *J. Exp. Med.* **169**, 779–794.
9. Gutstein, N. L., Seaman, W. T., Scott, J. H., and Wofsy, D. (1986) *J. Immunol.* **137**, 1127–1132.
10. Waldmann, H. (1989) *Annu. Rev. Immunol.* **7**, 407–444.
11. Carteron, N. L., Wofsy, D., and Seaman, W. (1988) *J. Immunol.* **140**, 713–716.

PART V

AUTOIMMUNITY

15

Activation-Induced Cell Death of Effector T Cells: A Third Mechanism of Immune Tolerance

CHARLES A. JANEWAY, JR., AND YANG LIU
Section of Immunobiology
Yale University School of Medicine
and Howard Hughes Medical Institute
New Haven, Connecticut 06510

INTRODUCTION

The fundamental characteristic of the immune system is its capacity to recognize and eliminate infectious agents while it discriminates these nonself components from self. Recent studies including several in this volume have provided numerous demonstrations of two major mechanisms of immune tolerance. First, clonal deletion has been shown to be an efficient mechanism by which self-reactive B or T lymphocytes can be eliminated from the immune repertoire on encountering antigen during their development (1, 2). For those lymphocytes which react with tissue-specific antigens, however, such clonal deletion is unlikely and indeed it has been difficult to demonstrate. In the latter situation, self-reactive T cells seem to have been rendered anergic by their interaction with tissue cells which lack costimulatory activity (3, 4). These two mechanisms, perhaps not mutually exclusive, seem to subserve different functions, to act in sequence, and to provide protection from autoreactivity and tissue destruction. However, these two mechanisms may not protect self-tissue from destruction by autoreactive cells which, through the activation by cross-reactive microbial infection (5), for example, have reached the effector cell stage. In this chapter we propose a third mechanism of

Molecular Mechanisms of
Immunological Self-Recognition

immune tolerance by which self-reactive effector T cells can be eliminated on encountering tissue cells which are devoid of costimulatory activity. Our proposal is based on an analysis of activation-induced death of cloned Th1 effector cells.

EXPERIMENTAL RESULTS

Most of the results to be presented were derived from clone 5.9, a Th1 clone which is specific for ovalbumin in association of I-A^d (6), but similar results have been obtained for five more Th1 clones with different antigen specificities. Clone 5.9 cells proliferate to cross-linked anti-CD3 monoclonal antibody (mAb) (YCD3-1, ref. 7) and anti-TCR Cβ mAb (H57-597, ref. 8) in the presence of irradiated T-depleted spleen adherent cells. In the absence of such accessory cells, anti-CD3/TCR mAbs fail to induce proliferation of clone 5.9. Interestingly, stimulation of clone 5.9 with anti-TCR/CD3 mAbs in the absence of accessory cells results in death of most of the T cells. Such cell death can be documented by trypan blue exclusion (60–90% of cell death after 16 hours of stimulation), fluorescence-activated cell sorting (FACS) analysis with propidium iodide, and loss of proliferation in response to recombinant interleukin-2 (IL-2) (90–99% inhibition).

To determine whether the surviving cells are rendered anergic by previous treatment with cross-linked anti-CD3 mAb, we took the approach of Jenkins and Schwartz (9) to determine the capacity of these survival cells to proliferate to antigen presented by spleen adherent cells. Our results demonstrate that on a cell-to-cell basis, the pretreatment with cross-linked anti-CD3 caused about 80% inhibition of proliferation to antigen. Thus, our results demonstrate that stimulation of cloned Th1 cells with cross-linked anti-CD3/TCR results in both cell death and clonal anergy. The relative contribution of each mechanism in functional inactivation of this Th1 clone is summarized in Fig. 1.

To analyze the mechanism of anti-CD3/TCR–induced cell death, cyclosporin A (CsA) was added to cultures of clone 5.9 cells. We found that CsA significantly inhibits cell death induced by anti-TCR/CD3 mAbs. As CsA is a potent inhibitor of cytokine production, we went on to test whether cytokines produced by clone 5.9 contribute to its own death. Our results demonstrate that anti–interferon-γ (IFN-γ) mAb (XMG 1.2, ref. 10) also totally blocks the cell death induced by anti-CD3. Anti–IL-2 mAb (S4B6.34, ref. 11), however, fails to inhibit cell death. Furthermore, recombinant murine IFN-γ can reconstitute the cell death inhibited by CsA. Thus our results, summarized in Table I, demonstrate that IFN-γ, a product of Th1 clones, is necessary for the activation-induced death of effector T cells.

Percent of Starting Response

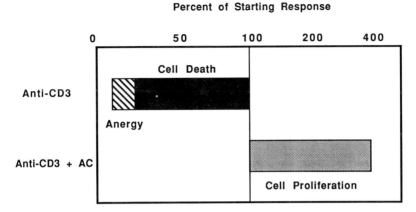

Fig. 1.   Consequences of anti-CD3 stimulation of Th1 clone in the presence or absence of accessory cells. Anti-CD3 alone killed about 75% of several Th1 clones tested within 16 hours; more than half of the surviving cells are anergic. When the same stimulus is applied in the presence of functional APCs, clonal expansion occurs.

INTERPRETATIONS

Clonal deletion is a successful mechanism for eliminating T cells whose receptors recognize self-antigens expressed by bone marrow–derived antigen-presenting cells (APCs); its utility in preventing T cell responses to self-tissue–specific antigens is unclear. The best data at present come from studies of T cell responses to artificially introduced tissue-specific antigens. Hammerling *et al.* have now provided some evidence which shows clonal deletion caused by tissue-specific antigens. In TCR–major histocompatibility complex (MHC) double transgenic mice with class I MHC antigen expressed only in brain, they observe significant deletion of T cells in the spleen but not in thymus (Hammerling *et al.*, this volume). In contrast, class I and class II MHC molecules expressed on pancreatic β cells do not cause deletion of potentially autoreactive T cells; instead, they lead to tolerance in the presence of potentially reactive cells (3, 4). This result is consistent with earlier studies of Jenkins and Schwartz (9), who showed that cloned T cell lines exposed to ligand in the presence of APCs underwent activation and clonal expansion, while in the absence of such APCs the T cells made effector lymphokines such as IFN-γ and lymphotoxin or tumor necrosis factor β (LT/TNF-β), but not interleukin-2. Moreover, such cells are resistant to further attempts to induce clonal expansion by antigens presented by potent costimulatory cells. As all cells except

TABLE I
IFN-γ Necessary for Activation-Induced Death of Th1 Clone 5.9

Treatment	Activation-induced cell death
Anti-CD3	++++
Anti-CD3 + Anti–IFN-γ	+/–
Anti-CD3 + CsA	+/–
Anti-CD3 + CsA + IFN-γ	++++

APCs (dendritic cells, macrophages, B cells) lack the costimulatory activity needed to induce clonal expansion and the achievement of effector function, encounters between naive T cells specific for a tissue antigen and the cell bearing that antigen lead to tolerance by clonal inactivation (12).

These two mechanisms, clonal deletion for ubiquitous antigens presented by APCs during intrathymic maturation and clonal anergy induced by tissue antigens expressed exclusively by noncostimulatory tissue cells, could provide good protection from autoreactivity and tissue destruction. However, these mechanisms cannot protect a tissue from already activated effector T cells which appear not to require costimulation to mediate their effector function. This is probably an important aspect of effector function directed against tissue cells; if effector T cells could not kill virus-infected tissue cells, even though these cells almost certainly lack costimulatory activity, then virus infections could not be cleared from the tissue. This scenario leaves the tissues vulnerable to certain immune effector responses. If a T cell recognizes a tissue antigen and is thus not deleted in the thymus, and if it has not yet encountered that antigen on a tissue cell, then activation of such cells by a cross-reacting environmental antigen could lead to a potent response directed against self-tissues. Is there any mechanism by which such tissues can be protected from sustained immunological attack?

Our study on activation-induced death of cloned effector T cells could suggest just such a third mechanism of immune tolerance. Thus, the fate of effector T cells is determined by costimulatory activity of the cells they react with; effector T cells can be further expanded if antigens are presented by cells with costimulatory activity. However, these same T cells are deleted if they interact with antigens expressed on cells that lack costimulatory activity. As the cell death we described requires interferon-γ produced by the effector cells, this mechanism is probably not used by naive T cells. We envisage that this tolerance mechanism is helpful in at least two circumstances: (1) inactivation of T cells which have reached effector cell stage by stimulation of cross-reactive microbes and (2) inactivation of T cells which were activated by conventional APCs that pick up tissue-specific antigens and drive them into effector cell stage. The basic characteristics of these different mechanisms of immune tolerance in relation to protection against autoimmunity to pancreatic β cells for example, are illustrated in Fig. 2.

Two very interesting findings reported in the symposium can now be interpreted in the light of this model. First, transfer of activated T cells specific for class I MHC antigen artificially introduced into mouse pancreas results in only

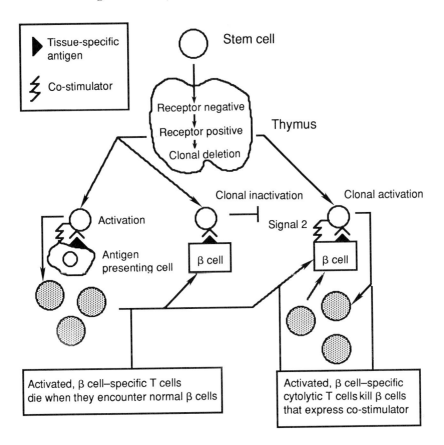

Fig. 2. Three mechanisms of immune tolerance: differential roles in the establishment and loss of tolerance to pancreatic β cells as a model tissue cell. Clonal deletion removes T cells specific for ubiquitous self-antigens such as self-MHC molecules. T cells binding to tissue-specific antigens on β cells are clonally anergized in the periphery (see refs. 3 and 4). T cells activated on costimulatory APCs could kill β cells, but the data suggest that such effector cells will die on binding β cells in the absence of costimulation. Only if β cells express costimulatory activity will sustained tissue damage result.

limited inflammation, which is rapidly resolved (Nossal, this volume). Second, when T cells which carry an H-Y–specific TCR are injected into irradiated nude male mice, such T cells rapidly expand. However, these activated effector cells are deleted shortly afterward (von Bohmer and Rocha, this volume). Thus, when T cells are transferred into male mice, conventional accessory cells will drive H-Y–specific T cells to expand and reach effector stage, but these effector cells apparently die when they encounter noncostimulatory tissue cells which carry the specific antigens. This raises a final and paradoxical question: What is the mechanism by which chronic inflammatory responses are sustained in tissues? The immune responses seem to be "set" to wage a limited war on invaders; what goes wrong to generate sustained conflict with host or engrafted tissues? This appears to be the central question for future development in this field.

ACKNOWLEDGMENTS

This work is supported by NIH grant AI-26810 to C.A.J., and Y.L. is a recipient of an Irvington postdoctoral fellowship.

REFERENCES

1. von Boehmer, H. (1990) *Annu. Rev. Immunol.* **8,** 531–556.
2. Nemazee, D. A., and Bürki, K. (1989) *Nature* **337,** 562–566.
3. Morahan, G. J., Allison, J., and Miller, J. F. A. P. (1989) *Nature* **339,** 622–624.
4. Burkly, L., Lo, D., Kanagawa, O., Brinster, R. L., and Flavell, R. L. (1989) *Nature* **342,** 562–564.
5. Oldstone, M. B. A. (1989) *Curr. Top. Microbiol. Immunol.* **145,** 1–141.
6. Conrad, P., and Janeway, C. A., Jr. (1984) *Immunogenetics* **20,** 312–319.
7. Portoles, P., Rojo, J., Golby, A., Bonneville, M., Gromkowski, S., Greenbaum, L., Janeway, C. A., Jr., and Bottomly, K. B. (1989) *J. Immunol.* **142,** 142–169.
8. Kubo, R. T., Born, W., Kappler, K. W., Marrack, P., and Pigeon, M. (1989) *J. Immunol.* **142,** 2736–2742.
9. Jenkins, M. K., and R. H. Schwartz. (1987) *J. Exp. Med.* **165,** 302–319.
10. Mosmann, T. R., Cherwinski, H., Bond, M. W., Giedlin, M. A. and Coffman, R. L. (1986) *J. Immunol.* **136,** 2348–2357.
11. Finkelman, F., Katona, I. M., Mosmann, T. R., and Coffman, R. L. (1988) *J. Immunol.* **140,** 1022–1027.
12. Schwartz, R. H. (1989) *Cell* **57,** 1073–1081.

16

T Cells Involved in Inductive Events in the Pathogenesis of Autoimmune Diabetes Mellitus

C. GARRISON FATHMAN, JAYNE S. DANSKA,
ALEXANDRA M. LIVINGSTONE,
AND JUDITH A. SHIZURU
Division of Immunology and Rheumatology
Department of Medicine
Stanford University
Stanford, California 94305

In a normal immune response, a complex of processed peptide presented in physical association with products of the major histocompatibility complex (MHC) on antigen-presenting cells (APCs) is recognized by the T cell antigen receptor (TCR) (1). The induction of both normal and autoimmune responses may be driven by the same types of inductive events. Activation of the CD4[+] T helper/inducer cell by TCR engagement with MHC class II–associated peptides causes lymphokine secretion, which in turn regulates effector cell activity (Fig. 1). Identification of the earliest (inductive) immunological events in the pathogenesis of autoimmune disease, particularly identification of the CD4[+] T helper/inducer cells bearing TCR specific for self-antigens, could provide potential sites of immunotherapeutic intervention. Toward this end, we have studied the inductive events in the pathogenesis of autoimmune insulin-dependent diabetes mellitus (IDDM).

Nonobese diabetic (NOD) mice spontaneously develop an autoimmune form of diabetes with many characteristics in common with human IDDM (2). Previous studies from our laboratory have demonstrated that removal of the CD4[+] T lymphocytes from the circulation using an anti-CD4 antibody (GK1.5)

Molecular Mechanisms of
Immunological Self-Recognition

Induction   Effectors

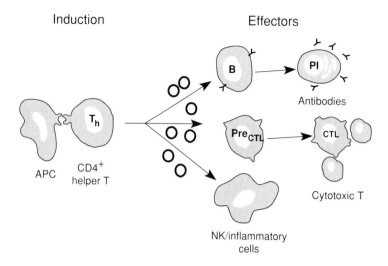

Fig. 1. Inductive events of immune response in which the CD4⁺ helper T cell recognizes antigen presented by the antigen-presenting cell (APC) in association with products of the major histocompatibility complex (MHC) class II genes. Following this recognition event, various lymphokines are produced by the CD4⁺ T inducer cells (o), which allows the differentiation of precursors into effector cells of the immune response. This includes B cells developing into antibody-secreting plasma cells, precursors of cytotoxic T lymphocytes developing into effector cytotoxic T lymphocytes, and a variety of other effector mechanisms generically listed on this diagram as NK (natural killer) cells or inflammatory cells.

successfully blocked the development of IDDM in NOD mice (3). These studies demonstrated that CD4⁺ helper/inducer lymphocytes are required for the development of autoimmune diabetes in NOD mice. In adoptive transfer experiments, other investigators have demonstrated that both CD4⁺ and CD8⁺ (effector) T cells are necessary for β islet cell destruction (4,5).

The hypothesis underlying our current work in NOD mice is that a relatively restricted set of TCR expressed on CD4⁺ cells is responsible for initiating the effector mechanisms which cause β cell dysfunction in IDDM. This suggestion is akin to the interpretation of data obtained in a model of demyelinating disease, experimental allergic encephalomyelitis (EAE) (6–9). In EAE, mice of susceptible MHC haplotype develop a demyelinating disease when immunized with certain antigenic fragments of myelin basic protein (MBP) (6). Following isolation of antigen-specific T cell clones from such mice, it was demonstrated that the clones capable of transferring disease to naive animals all expressed

$V_\beta 8$-containing T cell receptors (7). Furthermore, coadministration of MBP and monoclonal antibodies directed against these "pathogenic" T cell receptors blocked induction of EAE (8,9).

Several groups working on NOD mice have suggested that particular T cell receptor V_β gene families are used by lymphocytes which effect β cell destruction and hyperglycemia (10,11). The experimental basis of these claims has been either cloning of pathogenic T cells from inflamed islets or spleens of diabetic NOD mice or purposeful deletion of certain T cells expressing these TCR V_β region gene families in an attempt to block the development of diabetes (10–13). By these criteria, a variety of V_β gene families have been implicated, but none were demonstrated invariably to affect the development of NOD insulitis and/or diabetes (10–13).

We have attempted to ask whether certain TCR V_β genes are absolutely required for the development of IDDM in NOD mice by formal genetic analysis and by molecular and immunochemical examination of the expressed T cell receptor genes in the early islet infiltrates. To carry out the formal genetic analysis, we took advantage of the observation that certain strains of mice have a deletion of chromosome 6, which includes part of the TCR V_β gene cluster (14). Consequently, mice homozygous for $V_\beta{}^a$ lack V_β genes 5, 8, 9, 11, 12, and 13, which include several V_β families provisionally identified as being involved in the development of type I diabetes (10–13). NOD mice (wild-type $V_\beta{}^b$) were crossed with $V_\beta{}^a$ mice in order to obtain progeny that have the T cell receptor $V_\beta{}^a$ deletion on the background of diabetogenic genes from NOD. Backcross mice which maintained all of the appropriate NOD genes for the development of diabetes through two backcross generations were obtained (Fig. 2). Those mice were then intercrossed to obtain mice in which the T cell receptor haplotype segregated in a 1:2:1 fashion, giving rise to progeny of which one-quarter were homozygous for the $V_\beta{}^a$ T cell receptor chromosome 6 deletant haplotype, one-quarter were homozygous for the wild-type $V_\beta{}^b$ chromosomal segment, and one-half were $V_\beta{}^a/V_\beta{}^b$ heterozygotes. By analyzing these mice for the appearance of overt hyperglycemia by age 200 days and appearance of insulitis at 200 days, we were able to demonstrate that the T cell receptor V_β gene products contained within the $V_\beta{}^a$ deletion were not required for the development of diabetes or insulitis (15). Two of 16 female $V_\beta{}^a$ homozygotes developed diabetes with a mean day of onset indistinguishable from that of the littermate $V_\beta{}^{a/b}$ or $V_\beta{}^b$ homozygotes. Likewise, the development of insulitis was not correlated with T cell receptor V_β genotype in that mice of all three genotypes developed a similar incidence of the various grades of insulitis. These data clearly demonstrate that NOD mice of the $V_\beta{}^a$ genotype develop insulitis and diabetes, thereby proving that expression of the T cell receptor V_β gene products deleted in $V_\beta{}^a$ mice is not absolutely required for the development of disease.

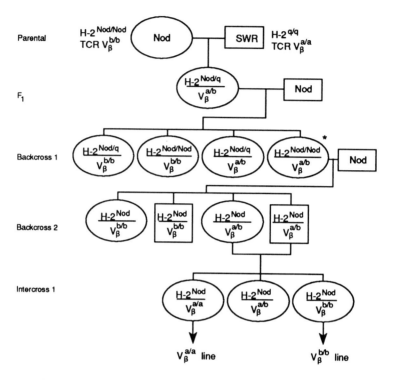

Fig. 2. The breeding scheme for the generation of T cell receptor (TCR) V_β heterozygous and homozygous mice. Asterisk (°) indicates that backcross 1 ♀ breeder had IDDM.

Our second approach to examining the TCR repertoire used by CD4⁺ T cells in IDDM induction has been to characterize TCR utilization of the first infiltrating T cells in the NOD pancreatic lesions. Such lesions occur many weeks before hyperglycemia. We identified the earliest time point at which we could observe lymphocytes infiltrating the islets of NOD mice in our colony. By histopathologic analysis, we were able to demonstrate that by 30 days of age, approximately 2 to 5% of the islets of about half of the female mice in our colony had detectable lymphocytic infiltrates (16). We reasoned that the cells infiltrating the islets in the earliest observable lesions might demonstrate oligoclonal or monoclonal utilization of T cell receptor specificities as was seen in the MBP-reactive T cells in the EAE model described above (8,9). To examine this question, we utilized techniques of polymerase chain reaction (PCR) amplification of cDNA synthesized from message contained within the isolated islets of mice at days 30, 40, and 50. Based on the formal genetic

TABLE I
TCR-V_β Representation in NOD[a]

Variable region	% of CD4	% of CD8	% total
2	4.6	1.1	5.8
	4.4	1.3	5.7
	4.8	1.0	5.8
	4.6	1.1	5.8
3	0.2	0.2	0.3–0.5
5	4.4	4.6	9.0
	3.4	6.8	10.2
	5.0	9.3	14.3
	4.3	6.9	11.3
6	5.8	10.0	15.8
	6.5	9.7	16.2
	5.8	12.6	18.4
	6.0	10.8	16.8
7	1.9	1.5	3.5
	1.6	1.9	2.5
	1.8	1.7	3.5
	1.8	1.7	3.1
8	5.3	5.2	10.5
	3.1	6.9	10.0
	4.5	5.3	9.9
	4.3	5.8	10.1
11	3.8	2.0	5.8
	3.8	2.2	6.0
	3.8	2.1	5.9
14	6.0	3.2	9.2
	6.2	2.5	8.7
	5.8	2.2	8.0
	6.0	2.6	8.6

[a]Peripheral lymph node cells from NOD animals at 60–75 days of age were placed in single cell suspension, stained with biotinylated anti-CD4 or CD8 and fluorescein isothiocyanate (FITC)–labeled reagents detecting each of the V_β determinants listed. FACS analysis was done in two colors. All numbers represent three or more animals and 20,000 cells counted. %CD4, 45–50; %CD8, 16–20.

analysis outlined above, we presumed the earliest infiltrating T cells would not express the T cell receptors contained within the deletant portion of chromosome 6 on the $V_\beta{}^a$ haplotype.

In order to determine which T cell receptors might be expected to be in the islets of NOD mice if there was random infiltration, we analyzed the peripheral T cell receptor V_β repertoire in lymph nodes of NOD mice at approximately 60 days of age. This analysis was done by cytofluorometric means, utilizing monoclonal antibodies directed at T cell receptor V_β determinants. These antibodies could detect $V_\beta 2$, $V_\beta 3$, $V_\beta 5$, $V_\beta 6$, $V_\beta 7$, $V_\beta 8$, $V_\beta 11$, and $V_\beta 14$. The V_β representation of such T cell receptors as the percent of CD4 and percent of CD8 or percent of total T cells is presented in Table I. These data suggest that NOD mice are of $V_\beta{}^b$ haplotype and probably possess the Mls 2^a allele, which deletes $V_\beta 3^+$ T cells during thymocyte maturation (17). By analyzing the expression of T cell receptor mRNA utilizing PCR amplification of cDNA from the lymph node lymphocytes of day 60 NOD mice, it was possible to identify, in a semiquantitative way, representation of the T cell receptor V_β gene products which had been

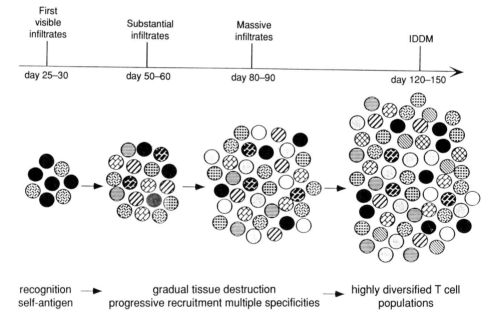

First visible infiltrates	Substantial infiltrates	Massive infiltrates		IDDM
day 25–30	day 50–60	day 80–90		day 120–150

recognition → gradual tissue destruction → highly diversified T cell
self-antigen   progressive recruitment multiple specificities   populations

Fig. 3. Diagram defining the progression to IDDM in our NOD colony in terms of lymphocytic infiltration of the pancreatic islets. It depicts our hypothesis that recognition of inductive autoantigens expressed in the β cells is a property of the earliest inducer T cells which enter the tissue. Such T cells express an extremely limited TCR repertoire. As inflammation proceeds, many T cells of divergent receptor specificities are recruited due to cellular destruction and lymphokine release, leading to a complex T cell repertoire within which the initial inducer subset, if present, is difficult to identify.

seen by cytofluorometric analysis. We then analyzed utilization of T cell receptor V_β family members by islet-infiltrating T lymphocytes at day 30 by PCR amplification of T cell receptor V_β gene mRNA (18). Controls for this study included (1) parallel examination of lymph nodes obtained from the same mice on the same day and (2) amplification of the thy-1 gene expressed in islet-infiltrating T cells.

Figure 3 is a representation of data obtained by such PCR analysis. The earliest infiltrating T cells, identified at day 30, express predominantly T cell receptor $V_\beta 3$ as assessed by PCR technology. By day 40, additional T cell receptor V_β gene products can be identified within the infiltrating islets, and by day 50 the V_β representation of the lymphocytes infiltrating the islets is as diversified as that expressed in lymph nodes of mice of the same age. These provisional data support our hypothesis that we may be able to identify a clonally restricted pool of T cells within the earliest infiltrating lymphocytes. Furthermore, these data support the idea that the induction of an autoimmune response may behave in a fashion analogous to that of a normal immune response (Fig. 1). Identification of early infiltrating T lymphocytes will allow a potential target of specific immunotherapy and may provide important tools for identification of the autoantigen expressed within the islets of Langerhans, the target of these particular T cell receptors.

ACKNOWLEDGMENTS

Supported by the Juvenile Diabetes Foundation International and NIH grants DK39959 and AI27989.

REFERENCES

1. R. H. Schwartz (1985) *Annu. Rev. Immunol.* **3**, 237–262.
2. Makino, S., Kunimoto, K., Muraoka, Y., Mizushima, Y., Katagiri, K., and Tochino, Y. (1980) *Exp. Anim.* **29**, 1–13.
3. Shizuru, J. S., Taylor-Edwards, C., Banks, B. A., Gregory, A. K., and Fathman, C. G. (1988) *Science* **240**, 659–662.
4. Bendelac, A., Carnaud, C., Boitard, C., and Bach, J. F. (1987) *J. Exp. Med.* **166**, 823–832.
5. Miller, B. J., Appel, M. C., O'Neil, J. J., and Wicker, L. S. (1988) *J. Immunol.* **140**, 52–58.
6. Zamvil, S., Nelson, P., Trotter, J., Mitchell, D., Knobler, R., Fritz, R., and Steinman, L. (1985) *Nature* **317**, 355–357.
7. Zamvil, S. S., Mitchell, D. J., Lee, N. E., Moore, A. C., Waldor, M. K., Sakai, K., Rothbard, J. B., McDevitt, H. O., Steinman, L., and Acha-Orbea, H. (1988) *J. Exp. Med.* **167**, 1586–1596.
8. Acha-Orbea, H., Mitchell, D. J., Timmerman, N. L., Wraith, D. C., Tausch, G. S., Waldor, M. K., Zamvil, S. S., McDevitt, H. O., and Steinman, L. (1988) *Cell* **54**, 263–273.

9. Acha-Orbea, H., Steinman, L., and McDevitt, H. O. (1989) *Annu. Rev. Immunol.* **7,** 371–405.
10. Reich, E.-P., Sherwin, R. S., Kanagawa, O., and Janeway, C. A. (1989) *Nature* **341,** 326–328.
11. Haskins, K., Portas, M., Bergman, B., Lafferty, K., and Bradley, B. (1989) *Proc. Natl. Acad. Sci. USA* **86,** 8000–8004.
12. Bacelj, A., Charlton, B., and Mandel, T. E. (1989) *Diabetes* **38,** 1492–1495.
13. Fukada, M., Horio, F., Kubo, R., and Hattori, M. (1989) *Diabetes* **38,** 12A.
14. Behlke, M. A., Chou, H. S., Huppi, K., and Loh, D. Y. (1986) *Proc. Natl. Acad. Sci. USA* **83,** 767–774.
15. Shizuru, J. A., Taylor-Edwards, C., Livingstone, A., and Fathman, C. G. (1991) *J. Exp. Med.* **174,** 633–638.
16. Danska, J. S., Ikeda, H., Taylor-Edwards, C., Weissman, I., and Fathman, C. G. (manuscript submitted).
17. Pullen, A., Marrack, P., and Kappler, J. W. (1988) *Nature* **355,** 796–801.
18. Guidos, C., Danska, J. S., Fathman, C. G., and **Weissman**, I. L. (1990) *J. Exp. Med.* **172,** 835–845.

17

Genetic Control of Diabetes and Insulitis in the Nonobese Diabetic Mouse: Analysis of the NOD.*H-2^b* and B10.*H-2^{nod}* Strains

LINDA S. WICKER[1], NICOLE H. DELARATO[1],
ALISON PRESSEY[1], AND LAURENCE B. PETERSON[2]

1. Department of Autoimmune Diseases Research
Merck Research Laboratories
Rahway, New Jersey 07065
and
2. Department of Cellular and Molecular Pharmacology
Merck Research Laboratories
Rahway, New Jersey 07065

INTRODUCTION

The nonobese diabetic (NOD) mouse is a model of spontaneous organ-specific autoimmunity (1). Nearly all individuals within this inbred strain of mouse develop insulitis, a lymphocytic infiltrate of the islets of Langerhans, and most become overtly diabetic. The infiltrates are predominantly composed of CD4$^+$ and CD8$^+$ T cells, and adoptive transfer of diabetes appears to require both of these subsets (2,3).

The breakdown of self-tolerance to the insulin-producing β cells within the islets is a consequence of the action of a number of genes in the NOD mouse, only one of which is linked to the major histocompatibility complex (MHC) (4–6). Estimates of non–MHC-linked NOD genes contributing to the

Molecular Mechanisms of
Immunological Self-Recognition

development of autoimmune diabetes range from three to six, depending on the inbred strain used for backcross analysis (7). In all analyses reported to date, only recessive diabetes-susceptibility genes (or dominant genes with low penetrance) would have contributed to the gene estimate. Fully dominant susceptibility genes would not be evident in F_1 × NOD first backcross (BC1) mice, since all BC1 mice would have at least one copy of the susceptibility alleles at all dominant loci.

Our efforts at understanding the various loci that contribute to disease development in the NOD mouse have focused on the recessive disease-associated loci segregating in the (NOD × B10)F_1 × NOD BC1 generation. In our initial studies, we found that an MHC-linked NOD gene functioned in an "almost" recessive manner, since some NOD/B10 MHC heterozygotes became diabetic, even though most diabetic mice were NOD MHC homozygotes (6). We went on to observe that a high percentage of NOD/B10 MHC heterozygotes develop insulitis despite the low incidence of diabetes (6,8). Importantly, we demonstrated that diabetic MHC heterozygotes do not represent a rare recombination event near the MHC but rather represent a low penetrance of the dominant MHC-linked diabetes-susceptibility allele in the NOD/B10 MHC heterozygotes (8).

The MHC-linked diabetogenic gene most likely represents two genes which code for the α and β chains of the class II MHC molecule of the NOD mouse, I-A^{nod}. The genes encoding the α and β chains of I-A^{nod} were sequenced by Acha-Orbea and McDevitt (9), and the β chain was found to be unique among inbred strains of mice. A likely role for the diabetogenic class II molecule is to present a β cell–derived peptide to the immune system that is not presented by nondiabetogenic class II molecules (10). In addition to this step, which is mediated by the MHC-linked diabetogenic gene, other non-MHC diabetogenic genes are responsible for the failure to prevent the generation of this autoimmune response. Candidate non-MHC diabetogenic genes include those responsible for proper T cell development, genes that control T cell signaling during tolerance induction and/or cell activation, genes which control cell adhesion and trafficking, and genes which control the expression of pancreatic antigens.

To understand the relationship of the MHC-linked and non–MHC-linked diabetogenic loci, we have developed two strains of mice: the NOD.*H-2^b* mouse, a strain which expresses all of the non–MHC-linked diabetes-susceptibility genes of the NOD but lacks the NOD MHC-linked diabetes-susceptibility gene, and the B10.*H-2^{nod}* mouse, a strain which expresses the NOD MHC-linked disease allele but not the non–MHC-linked disease alleles. The presence and absence of autoimmune phenomena in the NOD.*H-2^b* and B10.*H-2^{nod}* strains, respectively, emphasize the importance of the non–MHC-linked diabetogenic genes for developing diabetes in the NOD mouse.

MATERIALS AND METHODS

Animals

A breeding nucleus of inbred NOD mice was kindly provided by Dr. Yoshi-hiro Tochino (Aburabi Laboratories, Shionogi and Co., Osaka, Japan). Follow-ing cesarean derivation, our NOD colony has been maintained by brother–sister mating at Taconic Farms (Germantown, NY). NOD mice at Taconic Farms and Merck are housed under sterile, specific pathogen-free conditions. In our NOD colony, 80% of females and 50% of males develop dia-betes by 7 months of age.

To develop the NOD.H-2^b strain, NOD mice were outcrossed to C57BL/10SnJ (B10) mice obtained from The Jackson Laboratory (Bar Harbor, ME), and result-ing F_1 mice were backcrossed to the NOD strain as reported earlier (6,8). At each backcross, breeders were selected for the expression of I-A^b on the surface of pe-ripheral blood mononuclear cells. Generally, MHC heterozygous females were backcrossed to NOD males for each backcross generation. As described previ-ously (6,8), the prevalence of diabetes in NOD MHC homozygotes in the third backcross generation was equal to that of the parental NOD strain. At the eleventh backcross (N12) generation, an intercross was performed using male and female MHC heterozygotes. Intercross mice determined to be homozygous for K^b and I-A^b expression were used to initiate NOD.H-2^b inbreeding.

B10.H-2^{nod} mice were bred in a similar fashion except that (NOD x B10)F_1 mice were backcrossed to the B10 parent, and breeders for the next backcross were selected for expression of I-A^{nod} on the surface of peripheral blood mononuclear cells. Mice were intercrossed at the N6 generation to fix the NOD MHC on the B10 background.

MHC Analysis

MHC typing was performed using the monoclonal antibody 10-2.16 (Ameri-can Type Culture Collection, Rockville, MD), which reacts with I-A^{nod} but not I-A^b, followed by a counterstain, fluorescein isothiocyanate (FITC)–anti-mouse immunoglobulin G (IgG) (6), and the I-A^b–reactive monoclonal antibody TIB 120 (American Type Culture Collection, Rockville, MD) counterstained with FITC–anti-rat κ chain (AMAC, Westbrook, ME). All cells were analyzed by flow cytometry as previously described (6).

Assessment of Diabetes

Animals were monitored for elevations in urinary glucose using Tes-Tape (Eli Lilly, Indianapolis, IN) and were classified as diabetic after producing consistent Tes-Tape values of greater than 2^+. Mice classified as diabetic had tissues collected for histologic processing; pancreata were fixed in neutral-buffered formalin and processed for paraffin embedding. Tissue sections (4 μm) stained with H & E were examined to confirm that the pancreatic histology was consistent with the clinical observation of diabetes. Liver, kidney, heart, and spleen were frozen at −70°C as future sources of DNA.

Induction of Diabetes with Cyclophosphamide

Nondiabetic mice were treated with a two-dose regimen of cyclophosphamide to induce permanent clinical symptoms of diabetes (11,12). Cyclophosphamide (Cytoxan for injection, Mead Johnson Oncology Products, Evansville, IN) was prepared according to the manufacturer's guidelines immediately prior to dosing, and unused portions of cyclophosphamide were discarded. All mice received an initial intraperitoneal (i.p.) dose of 200 mg/kg and were tested twice weekly for elevated urinary glucose. Mice that had not become diabetic within 2 weeks received a second 200 mg/kg i.p. dose of cyclophosphamide, and monitoring for glycosuria was continued for an additional 2 weeks.

RESULTS AND DISCUSSION

Incidence of Diabetes and Insulitis in Parental and F_1 MHC Congenic Strains

Reciprocal transfer of the B10 and NOD MHC regions to the NOD and B10 backgrounds was accomplished by selective repetitive backcrosses. Table I summarizes the incidence of diabetes and insulitis in the MHC congenic strains NOD.H-2^b and B10.H-2^{nod} and F_1 mice involving these strains.

The NOD.H-2^b mouse only lacks expression of the diabetogenic MHC alleles. Thus, using this strain, contributions of the non–MHC-linked diabetogenic loci to the autoimmune status of the NOD can be assessed in the absence of the diabetogenic MHC. The NOD.H-2^b strain displays perivascular, periductal, and peri-islet lymphocytic infiltrates in the pancreas. Although these infiltrates in some individuals can become quite intense and involve large portions of the pancreas, widespread lymphocytic invasion of the islets (insulitis) or diabetes does not occur. Large lymphocytic accumulations are also observed in the submandibular glands of most NOD.H-2^b mice. Lymphocytic accumulations in both the pancreas and submandibular glands are predominantly comprised of T and B cells, as they are in the NOD mouse. Like the NOD, NOD.H-2^b mice also develop circulating autoantibodies to submandibular gland tissue (Michael

TABLE I
Autoimmune Status of Strains[a]

Strain	Mice with diabetes (%)	Mice with insulitis (%)	Pancreatic infiltrates
NOD	80 (females)	>95	Yes
	50 (males)		
NOD.H-2^b	0	0	Yes
(NOD × NOD.H-2^b) F_1	3 (females)	50	Yes
	0 (males)		
B10	0	0	Rare
B10.H-2^{nod}	0	0	Rare
(NOD × B10.H-2^{nod})F_1	0	0	Mild

[a]All nondiabetic mice were >7 months of age at the time of analysis. Insulitis was defined as mild to severe inflammation of at least one islet observed in two levels of pancreas. Pancreatic infiltrates consisted of infiltrating cells observed in periductal and/or perivascular locations as well as islet-associated inflammatory cells that did not appear to invade the islets.

Clare-Salzler, UCLA) and antibodies recognizing RINm38 rat insulinoma cells with a "polar" (13) staining pattern (Francesco Dotta, Rome). This latter result was unexpected because it was thought that the "polar" antibody might represent a specific marker for β cell autoimmunity. It is important to note that the B10 mouse, the donor of the MHC region of the NOD.H-2^b strain, does not display any of the autoimmune phenomena listed above. Thus, the NOD.H-2^b strain has been very valuable in helping to separate the non–MHC-linked diabetogenetic effects from the effects of the MHC-linked gene.

The (NOD × NOD.H-2^b)F_1 strain possesses diabetogenic alleles at all loci relevant to the development of diabetes but expresses only a single dose of the diabetogenic alleles at the MHC. This F_1 allows for the examination of the dominant versus recessive nature of the MHC-linked diabetes-producing gene. Although only 3% of (NOD × NOD.H-2^b)F_1 mice become diabetic spontaneously, mild to extensive insulitis is observed in approximately 50% of female F_1 mice (8) (Table I). Interestingly, the other 50% of F_1 females either display no pathology in the pancreas or have only infiltrates which are confined to the exterior of the islets in a perivascular or periductal manner. This wide variation in the pathology of F_1 mice, ranging from normal to severe widespread insulitis (which occasionally develops into diabetes), demonstrates that the NOD MHC is dominant with highly variable penetrance. This observed variability in the status of insulitis in the presence of one dose of the susceptible MHC suggests that the insulitis phenotype may not reliably differentiate between homozygous and heterozygous expression of a particular disease-susceptibility allele.

It is interesting to note that if insulitis had been the phenotype used to assess the contribution of the MHC in the BC1 generation, it is likely that the MHC would not have been associated with the development of insulitis because the MHC functions as a dominant gene for this phenotype. However, because of the lack of insulitis in the NOD.H-2^b strain (Table I) we know that the NOD MHC region is absolutely required for the development of the insulitis phenotype. In extending these observations to other non–MHC-linked loci which could also be dominant with low to moderate penetrance, only when resistance alleles from B10 are individually fixed in homozygous form on the NOD background can the relative contribution of each gene to the induction of insulitis and diabetes be assessed.

The B10.H-2^{nod} strain presumably carries only the MHC-linked diabetogenic gene of the NOD mouse. Individuals from the B10.H-2^{nod} strain, like those from the B10 strain, do not have abnormal accumulations of lymphocytes in their pancreata or submandibular glands. Thus, the MHC of the NOD mouse does not appear to cause autoimmune phenomena in the absence of the other non-MHC diabetogenic genes. An F_1 between NOD and B10.H-2^{nod} mice now allows the interaction of non–MHC-linked diabetogenic genes to be examined in an MHC homozygous environment. Previously we had examined only (NOD × B10)F_1 mice in which both MHC and non-MHC diabetogenic genes were in heterozygous form and found that both insulitis and diabetes were inherited in a recessive manner (6). Interestingly, like (NOD × B10)F_1 mice, (NOD × B10.H-2^{nod})F_1 mice have occasional perivascular and periductal infiltrates but are free of insulitis and diabetes. This indicates that the non–MHC-linked diabetogenic genes, when present in heterozygous form, act in a recessive fashion. This observation is in apparent conflict with our recent finding that each diabetes-susceptibility allele individually functions in a dominant manner (see Concluding Remarks). However, this conflict can be resolved by postulating that the lack of insulitis and diabetes in (NOD × B10.H-2^{nod})F_1 mice is due to the combined heterozygous expression of the non–MHC-linked diabetogenic loci.

Incidence of Diabetes following Treatment with Cyclophosphamide

It has been found that treatment of nondiabetic NOD mice with cyclophosphamide causes the majority of them to develop diabetes 7 to 14 days following treatment (11,12). In our colony, greater than 70% of nondiabetic female and male NOD mice ≥6 months of age develop diabetes after their first or second treatment with 200 mg/kg cyclophosphamide (unpublished observations, 1989) (Table II). In contrast to the diabetes observed in NOD mice, similar treatment of normal mice (including F_1 mice between NOD and normal mice) with cyclophosphamide does not cause diabetes (11,12). It has been postulated

TABLE II
Cyclophosphamide Induces Diabetes in NOD and
(NOD × NOD.H-2^b)F$_1$ Mice[a]

Strain	Sex	Number diabetic/ total number treated
(NOD x NOD.H-2^b) F$_1$	Female	11/41 (27%)
(NOD x NOD.H-2^b) F$_1$	Male	4/24 (17%)
NOD.H-2^b	Female	0/14
NOD.H-2^b	Male	0/5
NOD	Female	8/9 (89%)
NOD	Male	19/26 (73%)
B10.H-2^{nod}	Female	0/11
(NOD x B10.H-2^{nod})F$_1$	Female	0/15
(NOD x B10.H-2^{nod})F$_1$	Male	0/7

[a]Mice greater than 5 months of age received 200 mg/kg cyclophosphamide i.p. on days 0 and 14 and were observed for diabetes through day 28. All mice were nondiabetic at the initiation of cyclophosphamide treatment. It is important to note that the NOD mice represent animals that had not become spontaneously diabetic by 6 months of age. In our NOD colony, 75% and 80% of females and 48% and 50% of males exhibit spontaneous diabetes by 6 and 7 months of age, respectively. Thus, most of the NOD mice induced to become diabetic with cyclophosphamide would not have developed spontaneous diabetes within the time frame of this experiment.

that cyclophosphamide selectively removes suppressor T cells which can act to prevent the development of diabetes in the NOD mouse.

Treatment of the various strains revealed that only mice which spontaneously develop significant levels of insulitis, NOD and (NOD × NOD.H-2^b)F$_1$ mice, became diabetic following cyclophosphamide treatment (Table II). Mice that do not have spontaneous insulitis, NOD.H-2^b, B10.H-2^{nod}, and (NOD × B10.H-2^{nod})F$_1$, did not develop insulitis or become diabetic with cyclophosphamide. It is interesting that 89% of female and 73% of male nondiabetic NOD mice greater than 6 months of age developed diabetes following cyclophosphamide treatment, while less than 30% of (NOD × NOD.H-2^b)F$_1$ mice became diabetic with this treatment. We speculate that only those animals that have accumulated sufficient levels of insulitis, presumably due to elevated levels of β cell–specific T cells, become diabetic following cyclophosphamide treatment. Thus, the relatively low percentage of diabetes in (NOD × NOD.H-2^b)F$_1$ mice following cyclophosphamide treatment is consistent with the fact that only 50% of these mice possess mild to severe levels of insulitis, while 95% of NOD mice have moderate to severe insulitis (Table I). These observations suggest that cyclophosphamide acts to induce diabetes only following the spontaneous development of insulitis and does not itself induce insulitis and diabetes. In addition, the ability of cyclophosphamide to promote diabetes in (NOD × NOD.H-2^b)F$_1$ mice emphasizes the dominant nature of the NOD MHC.

CONCLUDING REMARKS

The NOD MHC congenic strains that we have developed will be valuable for characterization of the function of the MHC in the development of diabetes. For example, in our MHC heterozygous (NOD × NOD.H-2^b)F$_1$ mice that do not readily develop spontaneous diabetes, the transfer of NOD bone marrow to these mice results in a high incidence of diabetes (manuscript in preparation). Such observations indicate that heterozygous expression of the MHC in the thymus and pancreas is permissive for the development of diabetes in the NOD mouse and suggest further that homozygous expression of the MHC is required in the antigen-presenting cells that mature from the bone marrow.

Further definition of the non–MHC-linked NOD loci contributing to diabetes and insulitis is being carried out in collaboration with John Todd (John Radcliffe Hospital, Oxford) and Mark Lathrop (CEPH, Paris). By correlating the diabetes phenotype with polymorphic chromosomal markers throughout the mouse genome, one major finding is that the other non-MHC loci which contribute to the development of diabetes are (like the MHC-linked diabetes gene) dominant with varying levels of penetrance (manuscript in preparation). The occurrence of diabetic mice which are heterozygous at a particular diabetes-determining locus will make the fine mapping of the actual susceptibility gene more difficult because diabetics which have undergone recombination in the area of interest have the potential of being true recombinants between the marker locus and the actual disease gene or of being true heterozygotes which have become diabetic. Our efforts are currently focused on the development of NOD strains which express diabetes-resistance alleles (derived from the B10 strain) at each of the loci which influence the development of diabetes in this model. In addition, we are simultaneously developing B10 strains which express diabetes-susceptibility alleles (derived from the NOD strain) at each of the loci contributing to diabetes. Such strains will be useful in defining the activities of the diabetes-susceptibility allele at each of the contributing loci.

ACKNOWLEDGMENT

We would like to thank Dr. Nolan Sigal for his critical review of this manuscript.

REFERENCES

1. Makino, S., Kunimoto, K., Muraoka, Y., Mizushima, Y., Katagiri, K., and Tochino, Y. (1980). *Exp. Anim.* **29**, 1–13.
2. Bendelac, A., Carnaud, C., Boitard, C., and Bach, J.-F. (1987). *J. Exp. Med.* **166**, 823–832.
3. Miller, B. J., Appel, M. C., O'Neil, J. J., and Wicker, L. S. (1988). *J. Immunol.* **140**, 52–58.
4. Hattori, M., Buse, J. B., Jackson, R. A., Glimcher, L., Dorf, M. E., Minami, M., Makino, S., Moriwaki, K., Kuzuya, H., Imura, H., Strauss, W. M., Seidman, J. G., and Eisenbarth, G. S. (1986). *Science* **231**, 733–735.
5. Prochazka, M., Leiter, E. H., Serreze, D. V., and Coleman, D. L. (1987). *Science* **237**, 286–289.
6. Wicker, L. S., Miller, B. J., Coker, L. Z., McNally, S. E., Scott, S., Mullen, Y., and Appel, M. C. (1987). *J. Exp. Med.* **165**, 1639–1654.
7. Leiter, E. H. (1989). *FASEB J.* **3**, 2231–-2240.
8. Wicker, L. S., Miller, B. J., Fischer, P. A., Pressey, A., and Peterson, L. B. (1989). *J. Immunol.* **142**, 781–784.
9. Acha-Orbea, H., and McDevitt, H. O. (1987). *Proc. Natl. Acad. Sci. USA* **84**, 2435–2439.
10. Todd, J. A. (1990). *Immunol. Today* **11**, 122–129.
11. Harada, M., and Makino, S. (1989). *Diabetologia* **27**, 604–606.
12. Yasunami, R., and Bach, J.-F. (1988). *Eur. J. Immunol.* **18**, 481–484.
13. Dotta, F., Ziegler, R., O'Neil, J. J., Nayak, R. C., Eisenbarth, G. S., and Appel, M. C. (1989). *Diabetologia* **32**, 483A.

18

Islet Tolerance in Humans and Transgenic Mice

NORA SARVETNICK
Department of Neuropharmacology
Scripps Research Institute
La Jolla, California 92037

Recent work has demonstrated the utility of transgenic mice to further our understanding of immunological tolerance (1). The ability to target expression of antigens and immune effector molecules to the periphery *in vivo* provides a unique delivery system in which the acquisition of tolerance to "self"-components can be investigated. Although in most cases tolerance to self-components is maintained, in other rarer instances this tolerance is lost, leading to tissue destruction. We have been interested in autoimmune disease and specifically how in these states tolerance is lost and inflammation commences. We have specifically been investigating whether the response to a viral infection could lead to an autoimmune disease. To address this we have expressed an antiviral effector molecule, interferon-γ (IFN-γ) in the β cells of the islets of Langerhans in transgenic mice (2). These mice become diabetic following substantial pancreatic inflammation during the first few months after birth (Fig. 1). Immunochemical analysis has identified the infiltrating cells as T lymphocytes. The majority of these cells are L3T4-positive lymphocytes, with approximately 25% of these cells being Lyt2-positive cells.

We have been interested in determining whether these lymphocytes are capable of recognizing normal nontransgenic islet tissue *in vivo*. For this purpose we have transplanted histocompatible pancreatic tissue into the transgenic mice, assuming that if the animals had become sensitized to normal nontransgenic tissue they would reject the engrafted pancreas. This experiment showed that the mice had become sensitized to nontransgenic islet tissue *in vivo* (3). This experiment suggests that the localized expression of IFN-γ *in vivo* leads to the sensitization of the immune system. The exact mechanism of this recurrent inflammation in the histocompatible islet grafts is not known. To determine

Molecular Mechanisms of
Immunological Self-Recognition

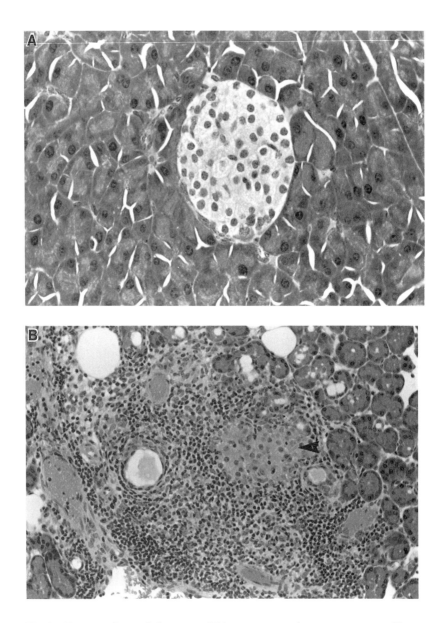

Fig. 1. Pancreatic histopathology in ins–IFN-γ transgenic and nontransgenic mice. The pancreas was fixed in formalin, embedded in paraffin, and stained with hematoxylin and eosin. (A) Section of a normal nontransgenic pancreas showing an islet. (B) Section of an ins–IFN-γ transgenic pancreas showing extensive inflammation in the peri-islet region.

whether the effect was specific for islet cells, histocompatible pituitary tissue was also grafted into the kidney of transgenic mice. This tissue was not rejected, indicating that the response of the transgenic mice was tissue specific with respect to these two tissues.

The grafting experiments could be interpreted as reflecting the loss of tolerance to islets in the IFN-γ transgenic mice or as reflecting the ability of low levels of this lymphokine in circulation to be deleterious to engrafted islets. In fact, several studies in the literature suggest that this lymphokine, in sufficient quantities, can be quite deleterious to islet cell function and viability (4). In order to distinguish between these possibilities, we performed breeding experiments to isolate the inflammatory component from any noninflammatory component of the induced transgenic disease. We decided to look at the effect of the lymphokine in the absence of the infiltrating T and B lymphocytes by backcrossing onto the immunodeficient SCID strain. We backcrossed the ins-IFN-γ transgenic mice to the SCID strain twice to obtain mice that were transgenic and also homozygous for the recessive mutant recombinase gene that SCID mice possess.

The IFN-γ/SCID transgenic mice had very distinct histopathology (3) from the immunocompetent version. The pancreata from these mice showed relatively normal islet cell mass and little inflammation. We therefore concluded that the lymphocytes caused the demise of the β cells in the transgenic mice. The islet morphology was relatively normal with the exception of a slightly "compressed" appearance of the cells within the islets. This type of "packed" appearance was reminiscent of the islet cells in the insulin-major histocompatibility complex (MHC) transgenic mice previously described (2). In addition, we assessed islet function by measuring the blood glucose of the IFN-γ/SCID transgenic mice. This revealed that the transgenic mice had mildly impaired islet function. Taken together, the morphological observations and the impaired islet function indicate that the expression of this lymphokine does indeed injure the β cells. Whether this injury necessarily precedes β cell loss via the lymphocytes is as yet unknown.

The results of these experiments need to be interpreted in the context of work performed by other investigators in the "transgenic immunology" area and with clinical data. Several recent reports have indicated that the expression of viral antigens in the β cells of transgenic mice does not lead to autoimmune phenomena (5,6). However, when these mice are infected with the intact virus, islet inflammation and diabetes rapidly result (5,6). This presumably occurs by a "molecular mimicry" pathway where the viral antigen initiates a cross-reacting response to the islet antigen. The response to the viral antigen occurs because this antigen is presented in the context of a virus which is more immunogenic, possibly due to the elicitation of cytokines. Taken together, this suggests that the presence of a "foreign" antigen in the β cells is not sufficient to initiate an autoimmune response *in vivo*. Instead, a state of nonresponsiveness results. If the antigen is presented in a more immunogenic form (viral infection), this "tolerance" is quickly lost and autoimmunity results. Since cytokines like IFN-γ are produced in response to virus infection, these are likely to be elicited when the viral infection occurs and could account for the loss of tolerance observed. In the ins–IFN-γ transgenic mice inflammation is induced which causes the trafficking of a large number of lymphocytes through the pancreas. When these lymphocytes encounter the IFN-γ in the transgenic islets, this cytokine could allow islet-specific T cells to respond to the islet antigens and lose their state of unresponsiveness.

In the clinical situation these studies are also relevant. Viruses infecting a genetically predisposed individual could cause loss of tolerance to viral antigens and by molecular mimicry to self-antigens. Insulin-dependent diabetes mellitus (IDDM) has recently been attributed to loss of immunological tolerance to the islet form of glutamic acid decarboxylase (7). This antigen has been shown to have some homology to coxsackie virus. Therefore a coxsackie infection could cause loss of tolerance to this islet antigen. It thus seems timely to surmise that viral infection and molecular mimicry can lead to IDDM in both humans and rodents. Additional disease models and clinical data on specific antigens are needed to determine if these correlations will hold true.

REFERENCES

1. Lo, D. (1990) *Current Topics in Microbiology and Immunology* **164,** 17.
2. Sarvetnick, N., *et al.* (1988) *Cell* **52,** 773.
3. Sarvetnick, N., *et al.* (1990) *Nature* **346,** 844.
4. Campbell, I. L. (1988) *J. Immunol.* **141,** 2325.
5. Oldstone, M. B. A., *et al.* (1991) *Cell* **65,** 319.
6. Ohashi, P., *et al.* (1991) *Cell* **65,** 305.
7. Baekkeskov, S., *et al.* (1991) *Nature* **347,** 151.

19

Fluorescence-Activated Cell Sorter Analysis of Peptide–Major Histocompatibility Complex

ANIMESH A. SINHA[1,*], TIMOTHY A. MIETZNER[2],
CHRISTOPHER B. LOCK[1], AND HUGH O. McDEVITT[1, 3]

1. Department of Microbiology and Immunology
Stanford University School of Medicine
Stanford, California 94305
and
2. Department of Molecular Genetics and Biochemistry
University of Pittsburgh School of Medicine
Pittsburgh, Pennsylvania 15261
and
3. Department of Microbiology and Immunology
and Department of Medicine
Stanford University
Stanford, California 94305

INTRODUCTION

In recent years it has become clear that the primary function of major histocompatibility complex (MHC) class I and class II molecules is to bind peptide fragments of protein antigens for their presentation to peptide-specific, MHC-restricted T cells. This is the basis by which self–nonself discrimination within the immune system is accomplished. The formation of MHC–peptide complexes is a prerequisite to T cell activation in normal and, presumably, autoimmune responses. Crystallographic data on human leukocyte antigen (HLA) class I molecules have revealed a peptide-binding groove, or cleft, that is formed by the membrane distal domains of the molecule (1–4).

*Current address: Department of Immunology, University of Alberta, Edmonton, Alberta T6H 5X4, Canada.

187

Although the association of peptides with MHC molecules has long been postulated, this phenomenon has been directly demonstrated only relatively recently. Peptide–Ia complexes were initially shown by equilibrium dialysis and subsequently isolated by gel filtration (5,6). Several other techniques have also been employed; these include chemical cross-linking methods as well as using radiolabeled peptides to detect cell surface binding (for review see ref. 7). Although many of these strategies are quite effective, they are in general difficult, require large amounts of purified Ia, are time consuming, and often require the use of radioactive labels. Using a method similar to that developed by J. Rothbard for human class II, we have used biotinylated peptides to analyze the interaction of peptides with class II MHC proteins at the surface of murine cells.

FACS BINDING ASSAY

Initially, two peptides were used in these binding studies: ovalbumin (OVA) 323–339 and hen egg lysozyme (HEL) 46–61 (Table I). These peptides elicit immune responses in reciprocal strains of mice. H-2^d mice are responders to the OVA peptide; H-2^k mice are nonresponders. In the case of the HEL peptide, H-2^k mice are responders and H-2^d mice are nonresponders. The peptides were biotinylated at their amino terminus with long-chain biotin (Sigma Chemical Co., St. Louis, MO) as detailed in Fig. 1. Neither peptide has any internal lysine residues that could serve as potential sites for internal biotinylation. Both biotinylated peptides were functionally active; each was shown to stimulate peptide-specific T cells *in vitro* and *in vivo* (data not shown).

Binding was first demonstrated using three murine B cell lymphoma lines: A20 (H-2^d), CH27 (H-2^k), and M12.C3 (class II negative). Class II expression at the cell surface was established by fluorescence-activated cell sorter (FACS) analysis using the allele-specific class II antibodies 10.3-6 (anti–H-2^d) and MKD6 (anti–H-2^k) (data not shown).

The binding protocol is outlined in Fig. 2. Briefly, biotinylated peptides are incubated with cells for a minimum of 1 hour at 37°C. Incubations at 40°C severely diminish binding. After washing, cultures are labeled with strepta-

TABLE I
Sequence and Immune Response Profile of Peptides Used in This Study

Peptide	Sequence	Responder	Nonresponder
OVA 323–339	ISQAVHAAHAEINEAGR	I-A^d	I-A^k
HEL 46–61	NTDGSTDYGILQINSR	I-A^k	I-A^d

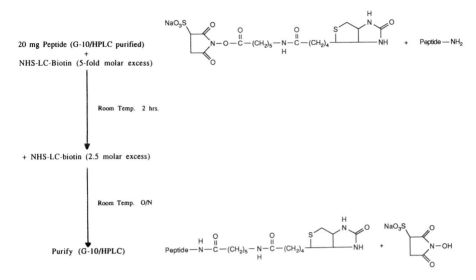

20 mg Peptide (G-10/HPLC purified)
+
NHS-LC-Biotin (5-fold molar excess)

Room Temp. 2 hrs.

+ NHS-LC-biotin (2.5 molar excess)

Room Temp. O/N

Purify (G-10/HPLC)

Fig. 1. Biotinylation of peptides. Long-chain biotin (Sigma Chemical Co.) was used for the biotinylation procedure to decrease the chance of disrupting the functional activity of the peptides. After biotinylation, the peptides were desalted over a G-10 column and purified by high-performance liquid chromotography. Both the OVA and HEL peptides were shown to stimulate specific *in vitro* and *in vivo* T cell responses.

vidin–Texas Red for approximately 30 minutes at 40°C, washed again, and resuspended in buffer for FACS analysis.

RESULTS

A fluorescent signal is produced and detected by FACS analysis only when the biotinylated peptides are incubated with cells expressing the responder class II allele (Fig. 3). OVA 323–339 binds to A20 (H-2^d) but not to CH27 (H-2^k) cells. In contrast, HEL 46–61 binds to CH27 but not to A20 cells (data not shown). As expected, no signal is seen when either peptide is incubated with M12.C3 (class II negative) cells.

Binding not only correlates with cells of the correct responder phenotype but also is clearly class II specific (Fig. 4). Binding of OVA 323–339 to A20 cells is inhibited by the appropriate allele-specific antibody MKD6 (anti–H-2^d) as well as by a 10-fold molar excess of native (unbiotinylated) OVA peptide.

Specific binding can also be shown using spleen cells, dendritic cells, and fibroblasts that have been transfected with class II molecules. We have also ex-

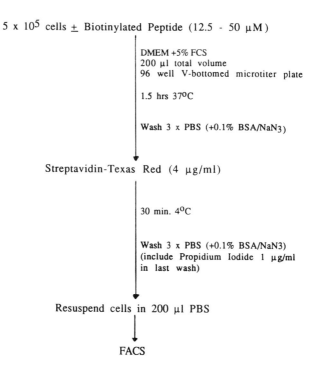

Fig. 2. FACS binding assay. Cells (1–5 million) are incubated with labeled peptide (10–100 μM) for a minimum of 1.5 hours at 37°C in 96-well V-bottom plates using DMEM tissue culture medium supplemented with 5% fetal bovine serum.

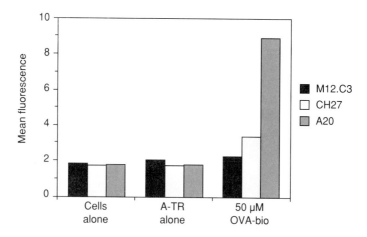

Fig. 3. Biotinylated OVA 323–339 (50 μM) binds to A20 (H-2^d) but not to CH27 (H-2^k) or M12.C3 (class II negative) following a 3-hour incubation at 37°C.

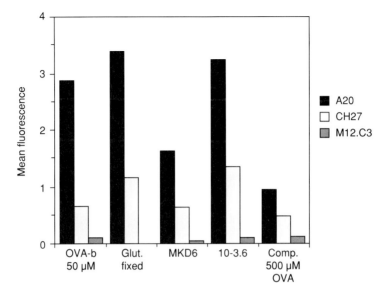

Fig. 4. Binding of OVA 323–339 to A20 is inhibited by coincubation for 6 hours with the antibody MKD6 (I-A^d specific) or a 10-fold excess (500 μM) of native unbiotinylated peptide but not by the antibody 10-3.6 (I-A^k specific) or by glutaraldehyde fixation of the cells for 30 seconds at .01%. These treatments had minimal effects on the binding to CH27 or M12.C3 cells.

tended this assay to other immunogenic peptides. Time course studies indicate a slow rate of association of peptide–MHC complexes that are similar to those determined by other methods.

DISCUSSION

In an earlier study, Berzofsky and colleagues showed that the presentation of a biotinylated myoglobin peptide to myoglobin-specific T cells could be blocked by adding avidin to the cell cultures (8). Rothbard and colleagues have developed a FACS binding assay to detect peptide binding to human class II molecules (9). We demonstrate here a simple, rapid, nonradioactive method for the detection of peptide–MHC class II complexes at the surface of murine cells.

Using FACS analysis, a large number of samples can be tested for binding in a very short period of time. Moreover, this method can be used to assess peptide–MHC binding on a variety of live cell types. The exact molecular nature of the association of peptides and MHC proteins remains to be fully elucidated. It also remains to be determined exactly which residues of Ia make contact with antigen and which make contact with the T cell receptor. To pinpoint these residues we are using the FACS binding assay in structure–function studies in which the ability of a large panel of cell lines expressing site-specifically mutated class II molecules to bind a peptide is compared with their ability to present this peptide to T cells.

ACKNOWLEDGMENTS

This work was supported by an Outstanding Investigator Award from the National Institutes of Health CA-49734. A.A.S. is the recipient of an Arthritis Foundation Investigator Award.

REFERENCES

1. Bjorkman, P. J., Saper, M. A., Samraovi, B., Bennet, W. S., Strominger, J. L., and Wiley, D. C. (1987) *Nature* **329,** 506.
2. Brown, H., Jardetzky, T., Saper, M. A., Samraovi, B., Bjorkman, P., and Wiley, D. C. (1988) *Nature* **332,** 845.
3. Garrett, T. P. J., Saper, M. A., Bjorkman, P. J. Strominger, J. L., and Wiley, D. C. (1991) *Nature* **342,** 692.
4. Saper, M. A. Bjorkman, P. J., and Wiley, D. C. (1991) *J. Mol. Biol.* **219,** 277.
5. Babbitt, B. P., Allen, P. M., Matsueda, G., Haber, E., and Unanue, E. (1985) *Nature* **317,** 357.
6. Buus, S., Sette, A., Colon, S. M., Miles, C., and Grey, H. M. (1987) *Science* **235,** 1353.
7. Rothbard, J. B., and Gefter, M. L. (1991) *Annu. Rev. Immunol.* **9,** 527.
8. Cease, K. S., Buckenmeyer, G., Berkower, I., York-Jolley, J., and Berzofsky, J. A. (1986) *J. Exp. Med.* **165**(5), 1440.
9. Busch, R., Strang, G., Howland, K., and Rothbard, J. (1990) *Int. Immunol.* **2,** 443.

20

Tolerance to Self: A Delicate Balance

ELLEN HEBER-KATZ
The Wistar Institute
Philadelphia, Pennsylvania 19104

INTRODUCTION

In natural populations, autoimmunity generally arises spontaneously and unpredictably. However, once established, the autoimmune state is usually irreversible in the sense that the acute/chronic course of the disease is often relentless, lasting a lifetime. As a consequence of the rarity of spontaneous appearance of disease, it has been necessary to study experimentally induced models of disease such as experimental allergic encephalomyelitis (EAE) as a model for multiple sclerosis and experimental allergic neuritis (EAN) as a model for Guillain–Barré syndrome. The autoantigen is presumably sequestered from the immune system in the normal state and has to be given in large doses with a potent adjuvant to induce the disease state. Antigen-specific T cells proliferate, find their target tissue, and mediate disease.

However, there is always the possibility that these models do not truly parallel the natural disease, a concern supported by the fact that most experimental models fail to achieve a chronic state. This chapter describes a chronic form of EAE which can be induced by immunization with synthetic T cell receptor (TCR) peptides derived from TCRs that dominate the T cell response in EAE. We propose that the regulation of a network mediated through variable (V) regions of TCRs leads to a chronic state of disease, whereas antigen itself leads to acute disease.

Molecular Mechanisms of
Immunological Self-Recognition

MONOIDIOTYPY

In the encephalomyelitis system in rats, antigens from the central nervous system such as myelin basic protein (MBP) are injected and cause paralysis. The disease has an acute phase followed by rapid and total recovery. It thus appears that some form of homeostatic down-regulation is occurring. Besides the potential involvement of adrenal control (1), there are experiments suggesting specificity in this regulation. First, after an acute episode of disease, the animal is specifically resistant to that disease. A disease-causing clonal T cell population also can induce a specific state of resistance when an attenuated form is injected. Since such a T cell population is related to what is believed to be all other T cells involved in disease induction through common antigen specificity, the TCR has been implicated as the target of regulation (2).

Of course, the basis for such interpretations depends on the demonstration that TCRs used by autoimmune T cells are restricted to a particular set of receptor genes. This has in fact been found to be true for the rat (3–5) and for the mouse (6–8), where T cell populations involved in EAE when induced by MBP share TCRs with a dominant V gene for both the α and the β chains used. Also, within a given strain of mouse (Pl/J, B10.Pl, and SJL) or rat (Lewis) and for a given antigenic determinant, the junctional regions of the TCR are shared as well. This finding lends support to the idea that the TCR is a target of regulation.

Further evidence suggesting the importance of common usage of V regions is the fact that T cells of the mouse and the rat involved in EAE, EAN, experimental uveoretinitis (EAU), and adjuvant arthritis use TCRs with the same V regions, $V_\alpha 2$ and $V_\beta 8$, even though the protein antigens recognized are different and the minimal determinants are not similar in primary structure and are non–cross-reactive (in the case of different EAE and EAN determinants). The junctional regions of these TCRs are unique within an antigen specificity group but different between groups. The relevance of the shared V regions has been attributed to the recognition of a second ligand other than antigen plus MHC (9,10).

TCR-SPECIFIC ANTIBODY MODULATION OF DISEASE

Considering the dominance of given V regions in pathogenic T cells able to transfer the autoimmune diseases mentioned above, it is not unreasonable that antibodies directed to such V regions would have an effect on disease. In the mouse, antibody to the TCR $V_\beta 8$ could be used not only to prevent EAE but also to reverse disease (6,7). In the rat, however, indications of up-regulation of EAE were seen upon injection of small doses of antibody, whereas larger doses could prevent disease (11,12). Thus, antibody directed to the antigen receptor of the disease-modulating T cell could affect the course of EAE. How this anti-body actually functions—whether it comes between a regulatory T cell which can recognize the effector TCR and its ligand, acts at the level of antigen recognition, or directly acts on the T cell carrying this receptor—is not known.

TCR PEPTIDE MODULATION

Further evidence for TCR involvement in regulation was shown by using synthetic peptides corresponding to the V and junctional regions of the relevant TCR used by the Lewis rat in the response to the encephalitogenic MBP determinant (13,14). Lewis rats immunized with these peptides in adjuvant were able to resist a challenge of autoantigen. This resistance was accompanied both by a proliferative T cell response to the peptide and to cells bearing these TCRs (such cells could also transfer protection) and by an antibody response to the peptide. From such results, it was concluded that regulatory cells specific for TCRs could be activated by peptides derived from these receptors and then down-regulate cells bearing these receptors.

With the same peptides, our laboratory obtained different though neverthe-less interesting results (15) (see Table I), results perhaps parallel to those obtained with anti-TCR antibody. Injection of the TCR peptides in adjuvant in Lewis rats did indeed induce a response, and the peptides again proved to be regulatory in nature. However, what resulted was up-regulation of EAE. Thus, when rats were first given TCR peptides in adjuvant and then challenged with the autoantigen, myelin basic protein, the animals displayed enhanced disease. TCR peptide alone did not induce disease. Enhancement was demonstrated in multiple ways:

1. Animals displayed early onset of disease, an effect which we believe is a qualitative difference rather than a quantitative difference. Early onset of this response can be equated with a secondary response to self, similar to a second set reaction in graft rejection when an animal has previously seen the alloantigen.

2. The severity of disease was greater in the animals that had received the TCR peptide.

TABLE I
Effects of Preinjection of TCR Peptides on Subsequent EAE Induction in Lewis Rats

System	Early onset/ enhancement	Chronic disease	No disease	No effect
$V_\beta 8$ peptide + CFA	9/12	6/12	1/12	2/12
$J_\alpha 39$ peptide + CFA	1/2	0/2	0/2	1/2
CFA	0/12	0/12	0/12	0/12

3. The disease becomes chronic in many of the animals as a result of TCR peptide priming.

Although enhancement was the dominant effect resulting from immunization with TCR peptides, we also found that there were animals that were not different from the CFA controls, and in one case an animal remained healthy, potentially indicating suppression (see Table I). Thus, there appears to be a variety of effects obtainable with such immunization, although enhancement appears to be the most likely outcome in our hands.

DISCUSSION

What does this mean? First, it is clear that TCR peptide immunization can modulate the immune response to a self-antigen in different ways depending on factors that are presently not well defined. In some instances, the same peptide will result in either suppression or enhancement. If down-regulation is mediated through activated regulatory T cells, then it is not unreasonable that elimination of such cells might result in up-regulation of an antiself-response. In fact, most of our results are null results: we have not found antibody; we have not been able to maintain a T cell response *in vitro* to these TCR peptides, though there is an initial round of stimulation. Thus, we propose that the regulatory T cell does see TCR peptide but it is tolerized rather than activated.

We would like to propose that there are different classes of interaction. First, the response to a nonself-antigen leads to an immunity that is clearly delimited and quickly subsides, presumably when the antigen disappears. The response to a self-antigen given in adjuvant also leads to a delimited, acute response which also diminishes presumably when the injected antigen disappears. In this case, the manifestation of the immune response is an acute disease state. However, when a TCR peptide followed by an autoantigen in adjuvant is injected, there is often a chronic response. In this case, one of the antigens, the TCR, which is represented on circulating T cells in the periphery of the animal and continually replenished, is thus not eliminated and can chronically restimulate the T cells. This leads to a response not unlike that seen in unrelenting human autoimmune disease, and we propose that it is not antigen but rather antigen receptor that is regulating this response.

REFERENCES

1. Mason, D., McPhee, I., and Antoni, F. (1990) *Immunology*. **70,** 1.
2. Ben-Nun, A., Wekerle, H., and Cohen, I. R. (1981) *Nature*. **292,** 60.
3. Burns, F., Li, X., Shen, N., Offner, H., Chou, Y. K., Vandenbark, A. A., and Heber-Katz. E. (1989) *J. Exp. Med.* **169,** 27.
4. Chluba, J., Steeg, C., Becker, A., Wekerle, H., and Epplen, J. T. (1989) *Eur. J. Immunol.* **19,** 279.

5.   Zhang, X-M, and Heber-Katz, E. (1992) *J. Immunol.* **148,** 746.
6.   Acha-Orbea, H., Mitchell, D. J., Timmerman, L., Wraith, D. C., Taich, G. S., Waldor, M. K., Zamvil, S., McDevitt, H., and Steinman, L. (1988) *Cell* **54,** 263.
7.   Urban, J., Kumar, V., Kono, D., Gomez, C., Horvath, S. J., Clayton, J., Ando, D. J., Sercarz, E. E., and Hood, L. (1988) *Cell* **54,** 577.
8.   Zamvil, S. and Steinman, L. (1990) *Annu. Rev. Immunol.* **8,** 579.
9.   Heber-Katz, E. and Acha-Orbea, H., (1989) *Immunol. Today.* **10,** 164.
10. Heber-Katz, E. (1990) *Clin. Immunol. Immunopathol.* **55,** 1.
11. Owhashi, M., and Heber-Katz, E. (1988) *J. Exp. Med.* **168,** 2153.
12. Heber-Katz, E. Owhashi, M., Happ, M. P., Burns, F., Shen, N., and Li, X. (1988) *Ann. N.Y. Acad. Sci.* **540,** 576.
13. Howell, M. D., Winters, S. T., Olee, T., Powell, H. C., Carlo, D. J., and Brostoff, S. E. (1989) *Science* **246,** 668.
14. Vandenbark, A. A., Hashim, G., and Offner, H. (1989) *Nature* **341,** 541.
15. Clark, L., Esch, T., Otvos, L., and Heber-Katz, E. (1991) *P. N. A. S.* **88,** 7219.

21

Prevention, Suppression, and Treatment of Experimental Autoimmune Encephalomyelitis with a Synthetic T Cell Receptor V Region Peptide

HALINA OFFNER[1,2], GEORGE HASHIM[3],
YUAN K. CHOU[1,2], DENNIS BOURDETTE[1,4],
AND ARTHUR A. VANDENBARK[1,2,5]

1. Department of Neurology
Oregon Health Sciences University
Portland, Oregon 97201
and
2. Neuroimmunology Laboratory
Veterans Affairs Medical Center
Portland, Oregon 97201
and
3. Departments of Microbiology and Surgery
Columbia University
New York, New York 10027
and
4. Neurology Service
Veterans Affairs Medical Center
Portland, Oregon 97201
and
5. Department of Microbiology and Immunology
Oregon Health Sciences University
Portland, Oregon 97201

Molecular Mechanisms of
Immunological Self-Recognition

INTRODUCTION

Immunization with myelin basic protein (BP) induces specific T helper lymphocytes that cause experimental autoimmune encephalomyelitis (EAE), a paralytic and inflammatory disease of the central nervous system (CNS). The encephalitogenic epitopes of BP are recognized in association with MHC class II molecules (1–4), and thus the encephalitogenic T cell specificities to discrete regions of BP differ widely among animal strains (5–9). In spite of distinct patterns of epitope/MHC recognition, encephalitogenic BP-reactive T cells in several rodent strains express preferentially the $V_\alpha 2 : V_\beta 8$ gene combination in their antigen receptors (10–15). In the Lewis rat, encephalitogenic $V_\alpha 2 : V_\beta 8$ (+) T cells recognize either the I-A–restricted 72–84 sequence of guinea pig (GP) or rat (Rt) BP, or the I-E–restricted 87–99 sequence conserved in all BP (16). In contrast, nonencephalitogenic T cells that recognized the I-A–restricted 43–67 sequence of GP–BP did not express either $V_\alpha 2$ or $V_\beta 8$ (16). The expression of common T cell receptor (TCR) V region sequences by the encephalitogenic T cells provided a rationale to use TCR peptides for idiotypic immunoregulation (for review see ref. 17). Such an approach would allow a more precise definition of T cell determinants and mechanisms than has been possible using whole-cell vaccination techniques that can protect against EAE and other T cell–mediated autoimmune diseases (17–19).

From an immortalized rat T cell clone that retained specificity for the major encephalitogenic determinant, our colleagues cloned and sequenced the rearranged TCR α and β chains and deduced the corresponding amino acid sequences (10). We identified the hypervariable and complementarity-determining regions (CDR), which have been modeled to contribute contact residues that interact with the antigen/MHC complex (20, 21) and thus might be good targets for specific immunoregulation. Moreover, we analyzed these sequences for amphipathic α helices (22) and Rothbard epitopes (23) to identify predicted antigenic regions. From these evaluations, we predicted that residues 39–59 of the $V_\beta 8$ sequence that included the second complementarity-determining region (CDR2) would be highly immunogenic for T cells (24).

To investigate the immunogenic and potential regulatory properties of this sequence, a synthetic peptide (TCR-$V_\beta 8$-39-59) was prepared. For comparison, two additional peptides that included the CDR1 and CDR3 regions of the $V_\beta 8$ sequence, and two corresponding peptides from the CDR1 and CDR2 regions of the $V_\beta 14$ sequence (25) were synthesized. The sequences of these peptides are shown in Table I. Both the TCR $V_\beta 14$ CDR1 and CDR2 peptides were predicted to be immunogenic, unlike either the TCR $V_\beta 8$ CDR1 or CDR3 peptide.

TABLE I
Sequences of Synthetic Peptides from the TCR

Region	Gene	Sequence
CDR1 (25-41)	$V_\beta 8$	K Q N N N H N N M Y W Y R Q D M G
		$\circ$ $\circ$ $\circ$ $\circ$
	$V_\beta 14$	T V K G T S N P N L Y W Y W G A P G
CDR2 (39-59)	$V_\beta 8$	D M G H G L R L I H Y S Y D V N S T E K G
		$\circ$ $\circ$ $\circ$ $\circ$ $\circ$
	$V_\beta 14$	A P G G T L Q Q L F Y S F N V G Q S G L V
CDR3 (93-102)	$V_\beta 8$	A S S D – S S N T E

ANIMAL STUDIES

Immunogenicity of the Peptides

Although the TCR peptides chosen for study represent autologous sequences, it seemed crucial that they be recognized immunologically in order for immunoregulation to occur. The ability of the TCR peptides to induce T cell and antibody responses was evaluated by injecting 50 µg of peptide in CFA and measuring delayed-type hypersensitivity (DTH) reactions (ear swelling) and specific antibodies (ELISA). As shown in Table II, both the TCR-V_β8-39-59 peptide and the TCR-V_β14-39-59 peptide induced strong and specific DTH reactions and antibodies. As predicted from the algorithms, the TCR-V_β14-24-40 peptide, but not the TCR-V_β8-25-41 peptide or the TCR-V_β8-93-101 peptide, induced specific DTH responses (data not shown). These data demonstrate clearly that autologous TCR sequences can be immunogenic *in vivo* and illustrate the usefulness of the algorithms in predicting potentially antigenic regions of the TCR sequence.

Prevention and Suppression of EAE with V_β8-39-59 Peptide

Consistent with previous experiments, injection of the V_β8-39-59 peptide in CFA 40 days prior to challenge with GP–BP induced complete protection against clinical EAE (Fig. 1). The dramatic protective effects of this protocol have been confirmed independently by Howell and associates, using CDR3 peptides from V_α2 and V_β8 sequences (26), and by Karpus and Swanborg (personal communication), who observed complete protection in 12 of 12 rats

TABLE II
Cellular and Humoral Immunity to the TCR Peptides[a]

	ΔDTH (ear swelling) mm/100		ΔAntibody (ELISA) O.D. units	
Immunogen	$V_\beta 8$	$V_\beta 14$	$V_\beta 8$	$V_\beta 14$
$V_\beta 8$ peptide	25 ± 3	2 ± 2	1.2 ± 0.4	0.02 ± 0.01
$V_\beta 14$ peptide	1 ± 1	40 ± 5	0.02 ± 0.01	1.7 ± 0.4

[a]DTH reactions were measured 48 hr after injection of 50 μg TCR peptide in saline, using a pressure-sensitive micrometer. Antibody responses were measured by ELISA, using 25 ng plated peptide antigen, 1:160 dilution of antiserum, and peroxidase-labeled second antibody.

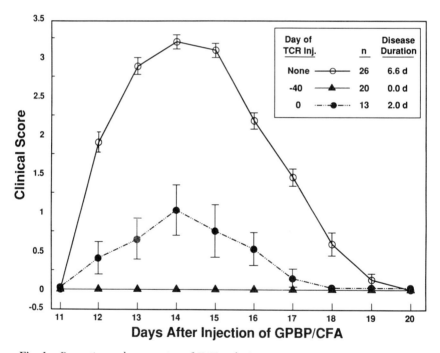

Fig. 1.   Prevention and suppression of EAE with 50 μg TCR $V_\beta 8$-39-59 peptide/CFA. Rats were injected s.q. with the TCR peptide 40 days prior to or at the same time as induction of EAE with 50 μg GP-BP/CFA. The clinical rating scale was as follows: 0, no signs; 1, limp tail; 2, hind leg weakness, ataxia; 3, hind quarter paralysis; 4, front and hind leg paralysis, moribund condition.

preimmunized with a TCR $V_\beta8$-39-59 peptide missing residues S-Y at positions 50–51 (see Table I).

Continuing with our own studies, simultaneous injection of the TCR-$V_\beta8$-39-59 peptide/CFA together with the encephalitogenic emulsion of GP–BP reduced the incidence (8 of 13 in the treated group versus 26 of 26 in controls), the maximum severity (score of 1.3 versus 3.4), and the duration (2.0 versus 6.6 days) of clinical signs of EAE (Fig. 1 and Table III). The results presented in Table III further show that $V_\beta14$-39-59 peptide, injected in CFA either simultaneously or 40 days before challenge with GP–BP, failed to alter the course of EAE in Lewis rats.

To avoid the use of CFA, we investigated the effects on EAE of administering a saline solution of the TCR-$V_\beta8$-39-59 peptide intradermally in the ear.

TABLE III
Prevention, Suppression, and Treatment of EAE with TCR $V_\beta8$-39-59 Peptide

	EAE			Clinical EAE	
Experimental group	Experiments	Total	Onset	Maximum clinical signs	Duration
GP–BP/CFA only	5	26/26	12	3.4 ± 0.2	6.6 ± 0.9
TCR/CFA					
Day –40	4	0/20	—	0 ± 0	0 ± 0
Day 0	3	8/13	13	1.3 ± 1.1^a	2.0 ± 1.8^a
Day 10	1	5/5	13	2.8 ± 0.4	3.6 ± 0.9^a
Day onset	4	20/20	12	2.3 ± 0.9^a	3.5 ± 1.4^a
Saline/DFA					
Day –40	2	8/8	12	3.3 ± 0.4	6.5 ± 0.9
Day 0	1	6/6	12	3.4 ± 0.3	6.6 ± 0.5
Day onset	1	6/6	12	3.3 ± 0.2	6.7 ± 0.8
$V_\beta14$/CFA					
Day –40	2	8/8	12	3.2 ± 0.3	6.1 ± 0.6
Day 0	1	6/6	12	3.3 ± 0.5	6.4 ± 0.3
Day onset	1	4/4	12	3.0 ± 0.4	6.0 ± 1.5
TCR i.d.					
Day 0	1	5/6	15	1.8 ± 1.0^a	3.0 ± 1.8^a
Day 7	1	4/6	14	1.7 ± 1.4^b	2.0 ± 1.9^a
Day before onset	1	6/8	12	1.8 ± 1.3^b	2.5 ± 2.1^a
10 μg, day of onset	1	6/6	12	2.8 ± 0.4	4.0 ± 0.9^a
50 μg, day of onset	1	9/9	12	2.8 ± 0.3	3.1 ± 0.3^a

[a] $p < .01$.
[b] $p < .05$.

The results show that the intradermal administration of 50 μg of the TCR peptide, given at the same time as the encephalitogenic challenge (day 0) or on days 7 or 11 after challenge, suppressed the maximum clinical severity from 3.4 to 1.7–1.8 and shortened the duration of EAE from 6.6 days to 3–4 days (Fig. 2 and Table III). These results indicate that the TCR peptide was effective in suppressing EAE, even when given in relatively low doses as late as only 1 day prior to the onset of EAE. Moreover, administration of the peptide i.d. in saline circumvented the irritating effects of CFA.

Correlation between Predicted Antigenicity of TCR Peptides and EAE Protection

In addition to the TCR-V_β8-39-59 peptide and the TCR-V_β14-39-59 peptide, other peptides from the CDR1 and CDR3 were tested for their ability to protect against EAE. The data from these experiments, summarized in Table IV, demonstrate that only antigenic regions from the relevant TCR (e.g., CDR2 from V_β8 sequence) have the capacity to protect against EAE. In contrast, nonantigenic regions of the relevant TCR (e.g., CDR1 and CDR3 of V_β8) or

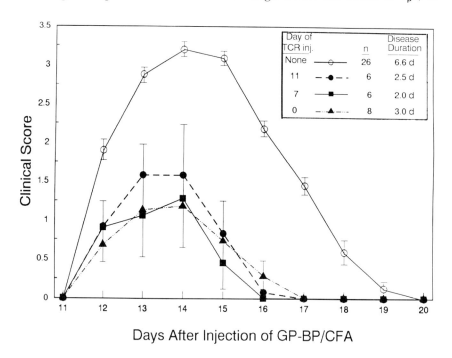

Fig. 2.   Suppression of EAE with 50 μg TCR V_β8-39-59 peptide given i.d. in the ear at the same time as or on days 7 or 11 after induction of EAE with 50 μg GP-BP/CFA.

TABLE IV
Immunogenicity and Protective Activity of TCR Peptides

		Predicted antigenicity	DTH induction	EAE protection
CDR1	$V_\beta 8$	− −	− −	− −
	$V_\beta 14$	+ +	+ +	− −
CDR2	$V_\beta 8$	+ +	+ +	+ +
	$V_\beta 14$	+ +	+ +	− −
CDR3	$V_\beta 8$	− −	±	±

antigenic regions of an irrelevant TCR (e.g., CDR1 and CDR2 of $V_\beta 14$) had little or no protective effect against EAE. Preliminary experiments (not shown) indicate that antigenic regions of the relevant $V_\beta 8$ sequence that do not include a CDR are also protective against EAE, suggesting that antigenicity of the peptide rather than its postulated antigen/MHC binding residues is the determining factor for activating EAE protective mechanisms.

CNS Histology in Clinically Well Rats

To evaluate the effects of TCR-$V_\beta 8$-39-59 peptide vaccination on histological EAE, brain and spinal cord sections from sick and clinically protected rats were compared for CNS lesions. After fixation, the sections were stained with hematoxylin and eosin. Comparable numbers of perivascular lesions were observed in the CNS from both control (sick) and TCR peptide-protected rats. A representative histological section from a TCR-protected, clinically well rat is shown in Fig. 3. These results demonstrate that fulminant CNS histology may occur in clinically unaffected rats and indicate that the presence of lesions in the CNS is not sufficient to account for the severe clinical deficits in EAE. Moreover, the presence of inflammatory lesions in the CNS of clinically well rats suggests that the effects of immunization with the TCR-$V_\beta 8$-39-59 peptide include mechanisms that regulate the function of encephalitogenic effector T cells in the tissues.

T Cell Responses in the CNS of Protected and Control Rats

The presence of mononuclear lesions in the brain and spinal cord of rats protected from EAE by TCR-$V_\beta 8$-39-59 peptide immunization raised interesting questions about the composition and function of T cells from the target organ.

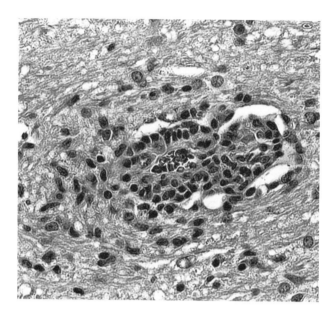

Fig. 3.   Paraffin-embedded section stained with hematoxylin–eosin from the spinal cord of a Lewis rat 20 days after immunization with GP–BP/CFA. This animal had been completely protected from clinical EAE by immunization with the TCR $V_{\beta}8$-39-59 peptide. This section reveals a perivascular infiltrate composed of mononuclear cells indistinguishable from that seen in control rats with severe clinical EAE. Magnification: ×300.

To evaluate possible differences in cell numbers, phenotypes, and antigen specificities, T cells were isolated and purified from fresh spinal cords collected from rats undergoing EAE or protected from clinical EAE by TCR peptide preimmunization. Figure 4 shows the scheme used for separation and testing of CNS-derived lymphocytes: Spinal cords were removed from groups of 4 rats, homogenized in medium, and separated by Percoll as described earlier (27, 28). The lymphocyte fraction obtained from the interface of the 40–50% Percoll layers was washed and the total cell number counted and phenotyped. Half of these lymphocytes were tested directly for responses to antigens. The remaining cells were cultured in interleukin-2 (IL-2) for 7–14 days prior to testing for antigen responsiveness. In our previous study (26), we found that activated T cells present in the CNS could be selectively expanded by IL-2 incubation, thus allowing us to distinguish cells that had penetrated the blood–brain barrier from those that were recruited nonspecifically into CNS lesions or that were not reactivated by specific antigen in the CNS.

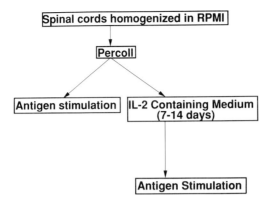

Fig. 4. Scheme for the isolation and testing of CNS-derived lymphocytes.

On average, approximately 1.2 million lymphocytes could be obtained per spinal cord from untreated control rats recovering from EAE 20 days after injection with GP–BP/CFA (Fig. 5). Similarly, 1.4 million cells per spinal cord were obtained from unprotected rats recovering from EAE after pretreatment with TCR-V_β14-39-59 peptide/CFA. Not surprisingly, almost the same number of lymphocytes, 1.2 million cells per spinal cord, were obtained from rats that had remained clinically well after a protective preimmunization with TCR-V_β8-39-59 peptide/CFA (Fig. 5). Phenotypically, the spinal cord lymphocyte populations from these three groups of rats were also quite similar, with no significant differences being detected in percentage or mean fluorescence intensity of staining for expression of T total (72–82%), T helper (65–75%), T suppressor/cytotoxic (18–20%), Ig (1–4%), macrophage (2–4%), I-A and I-E (60–70%), or IL-2 receptor (21–23%) markers. Thus, protection from clinical EAE by TCR-V_β8-39-59 peptide immunization did not alter the number of infiltrating cells or the predominant phenotype of activated T helper cells in the target organ. It is noteworthy that CNS cells from both protected and nonprotected rats expressed significantly higher levels of activation markers (MHC class II and IL-2 receptor molecules) than lymph node cells from the same rats (0–1% I-A and I-E and 6–9% IL-2R).

Spinal cord–derived T cells generally responded weakly when tested directly with specific antigens. As shown in Fig. 5, most antigen-driven proliferation responses were less than 3000 cpm per 500,000 cells, approximately 10- to 50-fold less than a comparable lymph node population (28). Qualitatively, the major differences in antigen responses between the unprotected, EAE-recovered donors

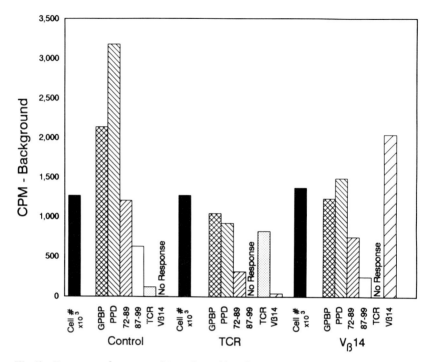

Fig. 5. Recovery and response of CNS-derived lymphocytes after direct antigen stimulation of 0.5 million cells from groups of 4 rats 20 days after the induction of EAE with GP–BP/CFA. The control group was not treated and developed severe EAE. The TCR group was preimmunized with 50 μg TCR $V_\beta 8$-39-59 peptide/CFA 40 days prior to challenge with GP–BP/CFA and was completely protected from clinical but not histological EAE. The $V_\beta 14$ group was preimmunized with 50 μg TCR $V_\beta 14$-39-59 peptide/CFA 40 days prior to challenge with GP–BP/CFA and developed clinical and histological signs of EAE identical to those of the control group.

and the EAE-protected donors were the markedly reduced response to the major encephalitogenic determinant of GP–BP (residues 72–89) and the total lack of response to the secondary encephalitogenic determinant (residues 87–99) in the TCR-$V_\beta 8$-39-59 peptide–immunized rats. As observed in lymph nodes, T cells specific for the respective TCR peptides were present in the spinal cord populations of TCR peptide–preimmunized animals (Fig. 5).

In contrast to the low responses observed on direct antigen stimulation, T cells expanded first in IL-2 had vigorous responses to antigens and mitogens, although the pattern of responses differed markedly. In the unprotected, EAE-recovered control groups, there were strong responses to Concanavalin A (Con A), GP–BP, and the encephalitogenic 72–89 determinant of BP, but no responses to PPD, the 87–99 determinant of BP, or the $V_\beta 14$-39-59 peptide even in $V_\beta 14$ peptide–immunized animals (Fig. 6). In the EAE-protected group, only relatively weak responses were observed to GP–BP and the 72–89 determinant, and no responses were present to Con A, PPD, 87–99, or TCR $V_\beta 8$ peptide.

Two conclusions can be drawn from these results. First, the presence of BP-reactive T cells but not PPD-reactive T cells after IL-2 expansion of CNS lymphocytes indicates that the BP-reactive cells were activated, whereas the PPD-reactive cells were not. These results suggest that the BP-reactive T cells had migrated from the periphery into the CNS and there remained activated after reencountering processed BP presented by accessory cells expressing MHC class II antigens. In contrast, PPD-reactive T cells from lymph nodes may have been activated initially, thus allowing passage across the blood–brain barrier, but failing reactivation in the CNS these cells may have reverted to an IL-2–nonresponsive state. Alternatively, the PPD-reactive T cells may represent a fraction of the nonspecifically recruited cells drawn into the CNS after alteration of the blood–brain barrier by activated BP-reactive T cells that

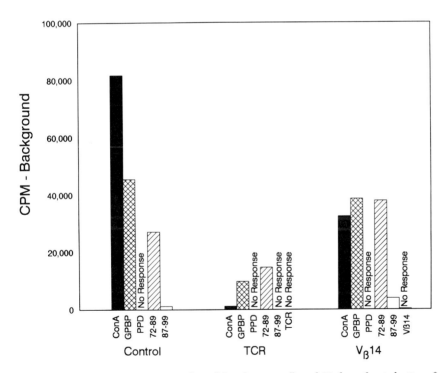

Fig. 6. Response of 2×10^4 CNS-derived lymphocytes collected 20 days after induction of EAE as in Fig. 5 and expanded for 7 days in IL-2 prior to antigen challenge. The groups indicated are the same as described in the legend for Fig. 5.

initiated the lesions. Second, the muted CNS responses to GP–BP and the 72–89 determinant of BP in TCR-V_β8-39-59 peptide preimmunized rats indicates a reduced encephalitogenic capacity and may explain in part why the rats were protected against clinical EAE.

In summary, although the total CNS cell number and phenotypes were essentially the same in TCR-V_β8 peptide–protected rats and unprotected control rats, the T cells present in the spinal cord of the protected rats were functionally anergic to the encephalitogenic epitopes of BP (Figs. 5 and 6). The number and size of the CNS lesions were similar in protected and unprotected rats, however (Fig. 3). Taken together, these data support the idea that there are activated BP-reactive T cells in the lesions that have altered encephalitogenic function, perhaps as a result of regulatory lymphokines such as transforming growth factor β (TGF-β), which can inhibit interferon-γ release without affecting IL-2 receptor expression (29, 30).

Proposed Mechanism

Immunication with the TCR-V_β8-39-59 peptide in CFA induced specific delayed hypersensitivity and antibody responses (Table II), indicating that both cellular and humoral arms of the immune response could be directed at this autologous TCR sequence. To evaluate their respective roles in protection against EAE, TCR-V_β8-39-59 peptide–specific T cell lines (24) and antibodies (29) were isolated and characterized. In the following section, the results of these previous studies will be summarized and a working model of the protective mechanisms proposed.

T cell lines specific for the TCR-V_β8-39-59 peptide could be selected easily from the lymph nodes of preimmunized rats, using the same techniques as for the selection of BP-specific lines (5). The T cell line was strongly CD4+ and weakly CD8+, and the response to the TCR-V_β8-39-59 peptide was inhibited by antibodies to MHC class I and CD8 molecules, but not by antibodies to MHC class II and CD4 molecules (24) (Table V). Moreover, the TCR-V_β8-39-59 peptide–specific line proliferated in response to irradiated encephalitogenic T cells expressing TCR V_β8 molecules on their surface, but not TCR V_β8⁻ T cell lines. However, no cytotoxicity was observed when the TCR V_β8+ T cells were used as targets. After activation with the TCR-V_β8-39-59 peptide presented by syngeneic irradiated thymocytes, as few as 5 million T line cells could transfer complete protection against EAE induced by injection of BP/CFA (Table VI). T cells cultured from the lymph nodes draining the site of the encephalitogenic inoculum responded strongly to the TCR-V_β8-39-59 peptide but poorly to the 72–89 epitope of BP (24). Surprisingly, the draining lymph node cells had increased responses to secondary epitopes of BP, a phenomenon termed "epitope switching."

TABLE V
Characteristics of TCR V_β8-39-59 Peptide-Specific T Cell Line

Stimulant	Proliferation (CPM/1000)
Medium	2 ± 1
TCR V_β8 peptide	100 ± 6
+ OX-6 (anti I-A)	99 ± 8
+ OX-17 (anti I-E)	100 ± 4
+ OX-18 (anti Class I)	$\underline{31 \pm 5}$
+ W3/25 (anti CD4)	105 ± 7
+ OX-8 (anti CD8)	$\underline{22 \pm 3}$
TCR V_β14 peptide	2 ± 0

[a]T cell line was selected from lymph nodes draining the site of injection of 50 μg TCR V_β8-39-59 peptide/CFA. Response was measured from 20,000 T line cells to 50 μg TCR peptides presented by 10^6 irradiated syngeneic thymocytes for 3 days prior to assay by uptake of labeled thymidine. Antibodies were added at 2 μg per microtiter well. Underlined values indicate significant inhibition.

In addition to T cell responses, immunization of Lewis rats or rabbits with the TCR-V_β8-39-59 peptide/CFA induced high titers of TCR peptide–specific antibody (31). Purified immunoglobulin obtained from the serum of immunized animals stained V_β8+ T cells, but not normal thymocytes which contain >5% V_β8+ T cells (Fig. 7). The antibodies were not cytotoxic for V_β8+ T cell targets, even in the presence of added complement. However, six injections every other day of 7–10 mg purified IgG starting at the same time as EAE induction reduced clinical signs of EAE from severe paralysis to very mild disease (flaccid tail) (31) (Fig. 8).

TABLE VI
Transfer of TCR Peptide–Specific T Cells Protects against Actively Induced EAE

	Induction of EAE (GP–BP/CFA)				DTH (mm/100)	
Transfer dose[a]	Incidence	Day of onset	Duration	Severity[b]	GP–BP	PPD
None	5/5	12	6.5	3.1	32	21
3×10^7	0/5	—	—	0	24	21
1×10^7	0/3	—	—	0	ND	ND

[a]T line cells were stimulated with TCR peptide + thymic accessory cells for 3 days prior to transfer i.p. into naive recipient rats. The rats were challenged on the same day with GP–BP/CFA.

[b]Value represents the mean of the maximum severity of EAE. 0, no signs; 0.5, lethargy, weight loss; 1, limp tail; 2, hind leg weakness; 3, hind quarter paralysis, incontinence; 4, moribund.

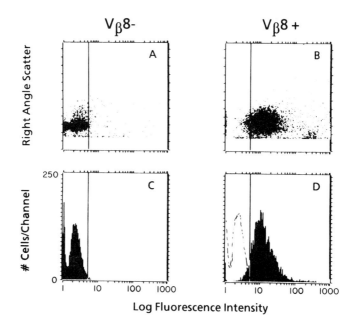

Fig. 7. Antibodies to TCR-V_β8-39-59 stain V_β8$^+$ encephalitogenic T cells. Normal thymocytes (A,C) or GP-S49S–specific T line cells (B,D) were incubated with rabbit anti–TCR-V_β8-39-59 antibodies, followed by a mouse antirabbit IgG facilitating antibody and fluorescein-labeled goat antimouse IgG antibody. Analysis was carried out on a Coulter Epics C Cytofluorograph. (A,C) Dot-plots of 10,000 cells showing cell size versus fluorescence intensity; (B,D) the corresponding histograms. Note the increased fluorescence intensity of anti–TCR-V_β8-39-59 antibody-stained T line cells (>90% stained) versus the normal thymocytes (5% stained). Both thymocytes and T line cells incubated with normal rabbit IgG instead of the anti–TCR-V_β8-39-59 IgG had background levels of staining (*dotted line in box D*).

These results demonstrate that both cellular and humoral mechanisms are capable independently of regulating V_β8$^+$ encephalitogenic T cell activity, and in preventing and suppressing EAE. Our working model of the mechanisms involved is presented in Fig. 9. The TCR-V_β8-39-59 peptide–reactive T cells home to the site of sensitization of BP-reactive T cells, where they down-regulate the induction of V_β8$^+$ T cells specific for the immunodominant 72–89 epitope of BP. This immunoregulation appears to be mediated by the regulatory T cells' ability to recognize a "processed" form of the TCR sequence presented in association with MHC class I molecules. The TCR peptide/MHC complex would constitute a legitimate T cell epitope for the anti-idiotypic T cells, in contrast to the intact TCR, which is not associated with MHC.

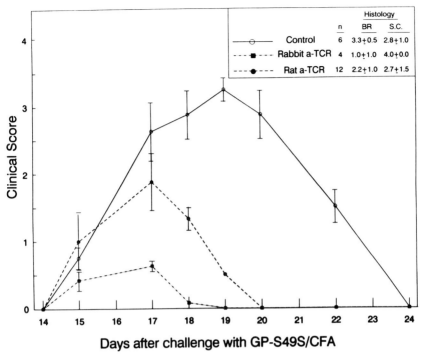

	Histology		
	n	BR	S.C.
Control	6	3.3±0.5	2.8±1.0
Rabbit a-TCR	4	1.0±1.0	4.0±0.0
Rat a-TCR	12	2.2±1.0	2.7±1.5

Days after challenge with GP-S49S/CFA

Fig. 8. Suppression of EAE with antibodies to TCR $V_\beta 8$-39-59 peptide. Lewis rats were challenged with encephalitogenic emulsions containing GP-S49S and 100 µg *M. butyricum* in the footpad. On the day of the challenge and every other day for 12 days, each rat received either 7 mg Lewis rat or 10 mg rabbit IgG from animals that had been immunized with TCR $V_\beta 8$-39-59. The IgG was dissolved in sterile saline and injected i.p. Treatment with normal rat and rabbit IgG did not influence the course of EAE in control rats. Animals were scored as described in Fig. 1.

The "processing" of the TCR sequence could take place in the Golgi before assembly of the TCR heterodimer on the T cell surface, or perhaps after the TCR complex had recycled off the cell surface. In either case, one would expect that the processed natural form of the TCR-$V_\beta 8$-39-59 peptide would be expressed at low density on the surface of the T cell, in competition with processed peptides derived from other molecules expressed by the T cell. Interaction of the TCR peptide/MHC I complex with the TCR/CD8 complex of the anti-idiotypic regulatory T cell presumably could trigger the release of inhibitory lymphokines such as TGF-β by the regulatory cells, or could inhibit directly the activity of the encephalitogenic $V_\beta 8^+$ T cell by perturbing associated membrane domains. Conceivably, the antibodies specific for TCR-$V_\beta 8$-39-59

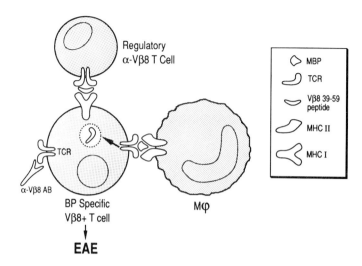

Fig. 9. Working model of anti-TCR V_β8-39-59 T cells and antibodies with encephalitogenic BP-specific V_β8$^+$ T cells. The EAE-inducing T cell expresses a processed form of the V_β8 molecule in association with MHC class I molecules that constitute a legitimate T cell target for the regulatory anti-V_β8 T cell. Antibodies to the TCR V_β8-39-59 peptide may react with either the intact or processed forms of the TCR V region.

peptide could react with exposed regions of the intact TCR or with a processed form of the TCR peptide associated with MHC.

Treatment of Established EAE with TCR V_β8-39-59 Peptide

The studies described above have demonstrated the effects on EAE of preimmunizing with the TCR peptide. To evaluate its therapeutic effects, the TCR V_β8-39-59 peptide was injected into rats on the first day of onset of clinical signs of EAE. These rats exhibited hind leg weakness, ataxia, and incontinence, an average clinical grade of 1.8–1.9. When the TCR V_β8-39-59 peptide was injected intradermally, no clinical effect on EAE was observed during the first day. However, during subsequent days, the severity of EAE was reduced markedly versus controls (Fig. 10A). The 50 μg/rat dose of TCR peptide caused a faster resolution of EAE (3.1 days) than the lower 10 μg/rat dose of peptide (4.0 days), compared to 6.6 days for the controls. Similarly, treatment with the TCR peptide/CFA prevented further progression of clinical signs and shortened the duration of EAE from 6.6 days (controls) to 3.5 days (Fig. 10B). In

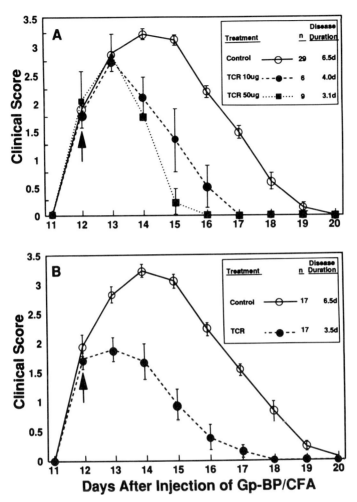

Fig. 10.   (A) Treatment of EAE with 10 or 50 μg TCR V_β8-39-59 peptide in saline given i.d. in the ear at onset of clinical signs (day 12) after induction of EAE with 50 μg GP–BP/CFA. Differences were significant between treated and control groups on days 14–18, as evaluated by Student's unpaired t-test.

   (B) Treatment of EAE with 50 μg TCR V_β8-39-59 peptide in CFA given s.q. at the onset of clinical signs (day 12) after induction of EAE with 50 μg GP–BP/CFA. Differences were significant between treated and control groups on days 13–18, as evaluated by Student's paired t-test.

contrast, treatment with TCR V_β14-39-59/CFA or CFA alone had no effect on clinical signs of EAE (Table III).

T Cell Responses in Rats Treated with TCR V_β8-39-59 Peptide

The rapid and dramatic clinical effect on EAE of the TCR-V_β8-39-59 peptide suggested that treatment with the peptide may have triggered a recall response to the TCR that in turn could regulate encephalitogenic T cell responses. To establish an immunologic basis for the rapid clinical changes, lymph nodes draining the site of the GP–BP/CFA challenge were evaluated in control and treated groups for T cell responses to the encephalitogen and TCR peptides on day 20, after recovery from EAE. As shown in Fig. 11A, the lymph node cells from rats treated with TCR-V_β8-39-59 peptide/CFA or TCR-V_β8-39-59 peptide in saline had modest responses to GP–BP and GP–BP peptides, similar to the untreated control rats. Of importance, however, lymph node cells from the treated groups had no response to Rt-BP, which must be recognized for EAE to occur. The TCR-treated groups also responded well to the TCR-V_β8-39-59 peptide. Surprisingly, lymph node cells from the control group also showed a modest but significant response to the TCR-V_β8-39-59 peptide, even though these cells had never before been exposed to this TCR peptide.

T cell responses to these antigens were amplified further by producing short-term T cell lines specific for either GP–BP or the TCR-V_β8-39-59 peptide. T cell line responses to BP determinants, although somewhat reduced, were clearly evident in both treatment groups (Fig. 11B), indicating that deletion of encephalitogenic clones did not occur. Moreover, both treatment groups had vigorous responses to the TCR-V_β8-39-59 peptide (Fig. 11C). It is noteworthy that T cell responses to the TCR-V_β8-39-59 peptide could also be amplified, albeit to a lesser degree, in the EAE-recovered control group (Fig. 11C). The persistence of BP responses and the strong T cell recognition of TCR-V_β8-39-59 peptide in the lymph nodes draining the site of the encephalitogenic challenge in treated rats are similar to results published previously of rats preimmunized with the TCR-V_β8-39-59 peptide in CFA 40 days prior to challenge with GP-BP (24). These results indicate that similar regulatory mechanisms were operative in both preimmunized and treated rats.

Induction of EAE Stimulates an Anti–TCR-V_β8 Network Response in Rats with No Prior Exposure to the Synthetic TCR-V_β8-39-59 Peptide

The rapid clinical benefit of TCR-V_β8-39-59 peptide therapy in rats with EAE, coupled with strong T cell recognition of the TCR peptide in the treated, EAE-recovered rats, prompted us to evaluate anti–TCR-Vβ8-39-59 responses *in vivo* and *in vitro* in rats that had never been exposed to the TCR-V_β8-39-59

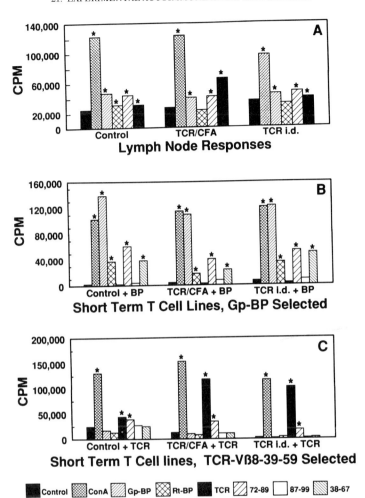

Fig. 11. (A) Proliferation response to GP–BP and TCR peptides of 5×10^5 lymph node cells collected after recovery from EAE from control and TCR $V_\beta 8$-39-59 peptide–treated rats.

(B) Proliferation response to GP–BP and TCR peptides of 2×10^4 short-term T line cells selected with GP–BP from LNC of control and TCR $V_\beta 8$-39-59 peptide–treated rats.

(C) Proliferation response to GP–BP and TCR peptides of 2×10^4 short-term T line cells selected with TCR $V_\beta 8$-39-59 peptide from LNC of control and TCR $V_\beta 8$-39-59 peptide-treated rats. Proliferation was measured by uptake of [^{3}H] thymidine over the last 18 hr of culture, and are expressed as average counts per minute of triplicate wells. Variation in replicate wells was less than 10%. An asterisk ($^\circ$) indicates that values that are significantly greater ($p < .05$ for paired samples) than the unstimulated control wells.

peptide. Thus, we measured ear swelling at the site of therapeutic i.d. administration of 50 μg TCR-V_β8-39-59 peptide in rats on the first day of clinical EAE induced with GP–BP/CFA. These treated rats had a significant DTH response to the TCR-V_β8-39-59 peptide (P < .01 compared to naive or CFA-immunized rats), but not to the V_β14 peptide or to an irrelevant peptide (LAM-1, Table VII) that appeared as the rats began their rapid recovery from EAE (Fig. 10). As expected, the magnitude of the DTH response to the TCR-V_β8-39-59 peptide in the recovering rats was understandably less than the response to the same TCR peptide in rats immunized 40 days previously with the respective TCR peptides in CFA, or than the response to GP–BP and PPD (Table VII).

Consistent with the DTH results, lymph node cells from untreated EAE-recovered rats contained T cells that proliferated *in vitro* in response to the TCR V_β8-39-59 peptide and to a lesser degree to the TCR V_β14-39-59 peptide (Fig. 12B), even though the donor rats had never been injected with the synthetic TCR peptides. In contrast, lymph node cells or thymocytes from naive rats did not respond to either of the TCR peptides. When lymph node cells from EAE-recovered rats were selected with the TCR V_β8-39-59 peptide, the specific response to this TCR peptide was amplified substantially to produce 40,000 cpm over background (Fig. 12B). Further selection of lymph node cells with the TCR V_β14 peptide amplified only V_β14-reactive T cells, but produced only 15,000 cpm over background. By comparison, preimmunization with CFA alone induced modest responses to both TCR peptides that could be amplified to produce 11,000 and 13,000 cpm over background (Fig. 12A). In contrast, selection of thymocytes or lymph nodes from naive rats with the TCR V_β8-39-59 peptide could not amplify a response to this peptide (data not shown).

TABLE VII
Transfer of TCR Peptide–Specific T Cells Protects against Actively Induced EAE

DTH test antigen	Immunization status				
	Naive	CFA	GP-BP/CFA[a]	V_β8/CFA	V_β14/CFA
TCR-V_β8-39-59	5 ± 2 (12)	6 ± 2 (4)	11 ± 2 (28)[b]	36 ± 8 (30)	6 ± 2 (9)
TCR-V_β14-39-59	8 ± 3 (15)	10 ± 4 (8)	8 ± 3 (6)	8 ± 2 (6)	40 ± 4 (12)
LAM/1-187-204	4 ± 2 (6)	4 ± 3 (6)	4 ± 3 (6)	4 ± 2 (6)	4 ± 1 (4)
GP–BP	7 ± 2 (6)	7 ± 3 (6)	36 ± 5 (6)	8 ± 3 (12)	8 ± 2 (4)
PPD	9 ± 3 (6)	34 ± 6 (6)	33 ± 4 (6)	35 ± 5 (6)	37 ± 4 (4)

[a]Rats injected with GP–BP/CFA were tested i.d. with 50 μg TCR-V_β8-39-59, TCR-V_β14-39-59, or LAM/1 (VRSKIGSTENLKHQPGGG) on the first day of clinical EAE. Values represent DTH (ear swelling) 48 hr after injection of the peptides.

[b]Underlined values indicate significant increase (p < .01) in ear swelling compared to naive rats 48 hr after injection of the test antigen, evaluated by Student's unpaired t-test. Parentheses indicate number of rats tested.

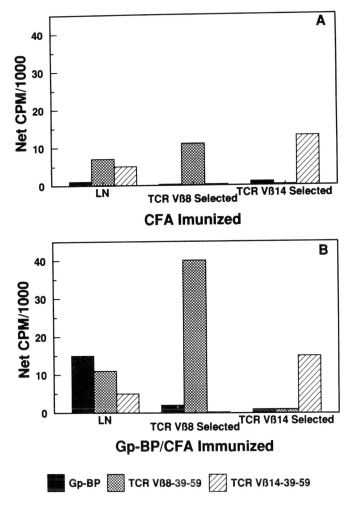

Fig. 12.   Response of lymph node cells and T cell lines to GP–BP, TCR V_β8-39-59 peptide, and TCR V_β14-39-59 peptide from rats immunized with CFA alone (A) or with GP–BP/CFA (B). Note responses to both TCR peptides in CFA-immunized LNC that can be expanded marginally into T cell lines. In contrast, note increased response to V_β8 peptide in GP–BP/CFA–immunized rats that is amplified strongly in the TCR V_β8-39-59 peptide–selected T cell line.

Thus, although CFA induced detectable responses to the TCR peptides, the inclusion of GP–BP in the inoculum resulted in a much stronger selective increase in the response to the TCR $V_\beta 8$-39-59 peptide.

The results presented above lend strong support to the idea that expansion of TCR $V_\beta 8^+$ encephalitogenic T cells during EAE induces a regulatory network directed in part at the TCR $V_\beta 8$-39-59 peptide. The ability of this TCR peptide to trigger a natural regulatory network and to produce a beneficial therapeutic effect is a major step forward in the quest to regulate selectively T cell responses to autoantigens in established disease. The rapid therapeutic effect of the TCR peptide is reminiscent of passive anti-TCR antibody therapy reported in the mouse and rat models of EAE (11, 32, 33). However, the TCR peptide would have several advantages over passive antibody therapy, including its lack of foreign antigenic determinants and its ability to trigger long-lasting natural regulatory mechanisms. In addition, the ability to treat EAE with an i.d. injection of peptide in saline (without adjuvants) provides a feasible route for the eventual treatment of humans.

Of equal importance is the demonstration that the regulatory network directed at the synthetic TCR peptide is induced as a consequence of the autoimmune disease itself. By implication, the mere presence of a given anti-TCR response in patients with autoimmune disease would indicate its importance as a target of the immune network that potentially could be triggered with clinical benefit by injecting small doses of the appropriate TCR peptide. Recent studies indicate that T cells from patients with multiple sclerosis, in which BP may be a relevant target autoantigen, utilize a limited set of V region genes in plaque tissue (34) and in response to human myelin basic protein (35, 36). Our data would predict increased antiidiotypic responses to the overexpressed TCR V genes if the T cells were involved in the pathogenesis of the disease, but no detectable or amplifiable responses if the T cells represent a baseline frequency. The presence of anti-TCR responses *in vitro* would thus provide the impetus for periodic TCR peptide therapy to boost the maintain specific antiidiotypic responses in patients with established autoimmune disease.

HUMAN STUDIES

Involvement of BP-Specific T Cells in Multiple Sclerosis

The spectrum of clinical and histological signs induced in a wide variety of animal species by T cells specific for basic protein or proteolipid protein resembles in many ways the human diseases multiple sclerosis (MS) and acute disseminated encephalomyelitis (ADE) (37, 38). Consequently, human T cell recognition of BP has been of considerable interest (35, 36, 39–48). Previously, we reported that MS patients' T cells responded to BP more frequently than controls (41), and the pattern of epitopes recognized suggested chronic sensitization to BP (46, 48). Other reports indicate that there is an increased frequency of activated BP-specific T cells in the blood of MS patients (36, 49), and a recent report by Link *et al.* (50) demonstrated a 40-fold increase in frequency of T cells from the cerebrospinal fluid of MS patients versus neurologic controls that secreted interferon-γ in response to BP stimulation. This study is of particular interest in view of the documented effect of IFN-γ to induce exacerbations in MS patients (51). However, the actual involvement of BP-reactive T cells in the pathogenesis of MS can be demonstrated only if selective regulation or removal of these cells can affect the disease process. As described above, such selective regulation is now possible using TCR peptide therapy, providing that human BP-reactive T cells also express common V region genes.

BP Epitope Specificity of T Cell Lines and Clones

To characterize BP-specific T cell responses in MS patients and controls, we repeatedly stimulated blood lymphocytes with whole human BP (Hu-BP) and then expanded the activated T cells with IL-2. As the T cell line developed, specific responses to immunodominant Hu-BP epitopes increased, in contrast to other common microbial antigens such as herpes simplex virus. T cell lines selected in this manner from immunized rats and mice are usually encephalitogenic, and the relevant immunodominant BP epitopes recognized by the line can be identified by mapping the pattern of T cell response to highly purified fragments and synthetic peptides of BP (16). Epitopes that are immunodominant and encephalitogenic in a given animal strain are influenced by the MHC haplotype expressed by the antigen-presenting cells used in the T cell line selection (1, 2). Thus, to ensure fidelity, only autologous blood monocytes were used to restimulate the human BP-specific lines and clones. In Caucasians of Northern European descent, HLA-DR2 is overrepresented in the MS patient population (60% versus 25% at large), and in the present study the proportion of patients and controls who were HLA-DR2 positive generally reflected the percentages in the respective populations (MS, 80% DR2+; controls, 30% DR2+).

Although the Hu-BP–specific T cell lines responded equally well to BP, Con A, and the C-terminal fragment of Hu-BP (P4), the MS patient–derived T cell lines had stronger and more frequent responses than normal donors to the 45–89 (P1) fragment and to a lesser degree the 1–38 (P2) fragment of Hu-BP (Fig. 13). T cells from 10 of the 11 MS lines responded to at least two of the three fragments of Hu-BP, whereas only 3 of 9 normal T cell lines recognized multiple Hu-BP epitopes. As reported earlier (45), T cell responses to Hu-BP were restricted predominantly by HLA-DR.

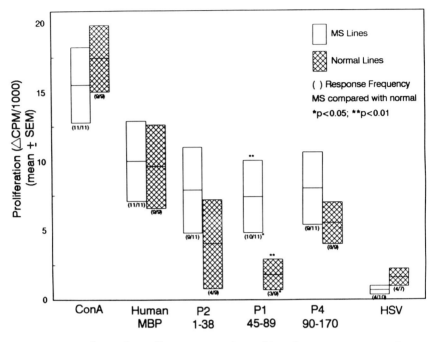

Fig. 13. Peptide specificity of human BP-specific T cell lines from 11 MS patients and 9 normal donors. Twenty million blood T lymphocytes were alternately stimulated with whole Hu-BP and IL-2 for 7–8 weekly cycles. Proliferation responses of 2×10^4 T cells were evaluated to the indicated antigens presented by 2×10^5 autologous irradiated (4500 rad) blood mononuclear cells.

To confirm the pattern of T cell line responses, Hu-BP–specific T cell clones were isolated and characterized from lines that had undergone two rounds of restimulation with whole Hu-BP. Each clone was expanded to approximately 1 million cells and tested for response against a battery of Hu-BP fragments and peptides. As shown in Fig. 14, the frequency of T cell clones from MS patients and controls reflected closely the pattern of line responses, with 1–38 and 45–89 specific clones occurring at a higher frequency in MS patients. Clones recognizing the 90–170 fragment of Hu-BP occurred at a similar frequency, whereas clones responding to none of the fragments were more frequent in controls. Within the 90–170 C-terminal fragment (Fig. 15), MS clones responding to the 130–149 peptide were more frequent than control clones, whereas control clones specific for the 110–129 peptide were more frequent than MS clones. The frequency of MS and control clones specific for the 87–99 and 149–172 peptides did not differ.

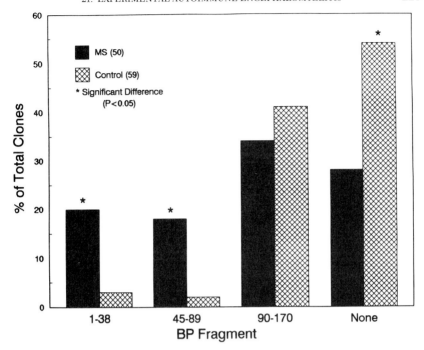

Fig. 14. Comparison of response patterns of 59 BP-specific T cell clones from 7 MS patients and 50 clones from 6 normal donors to peptic cleavage fragments of Hu-BP (kindly prepared by Dr. C.H.-J. Chou, Atlanta, GA). Bars indicate percentage of clones responsive to each peptide. Note the increased percentage of MS clones specific for both the 1–38 and 45–89 peptides and the increased percentage of normal clones that did not react with any BP fragment. Response to the 89–170 fragment did not differ between the two groups.

Taken together, these data demonstrate that circulating T cells from MS patients recognize more and different epitopes of BP than T cells from normal donors. In animals, the complexity of T cell responses to BP epitopes increased as a function of the time of sensitization (18). By inference, the more complex MS response to BP would result from chronic sensitization, probably linked to the process of demyelination. As a consequence, one might presume an increased risk of inducing encephalitogenic and demyelinating T cell specificities that could contribute to a destructive cycle of autoimmunity.

HLA-DR2 Is Capable of Restricting Multiple Human BP Epitopes

The association of the HLA-DR2/DQw1 haplotype with MS (52) suggests that these class II molecules, especially DR2, could play a critical role in restricting CD4+ T cell responses to CNS autoantigens. To evaluate epitopes

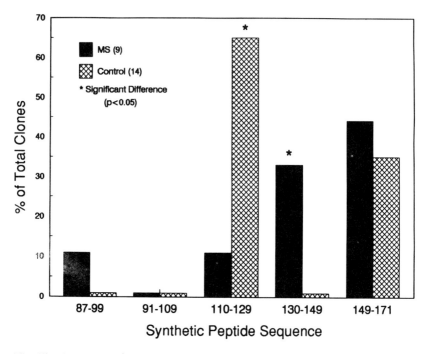

Fig. 15. Comparison of response patterns of nine 90–170–specific T cell clones from four MS patients and 14 clones from four normal donors to synthetic peptides within the 90–170 sequence of Hu-BP. Bars indicate percentage of clones responsive to each peptide. Note the increased percentage of MS clones specific for the 130–149 peptide and the increased percentage of normal clones specific for the 110–129 peptide. Response to the 149–171 peptide did not differ between the two groups.

restricted by DR2, the Hu-BP epitopes recognized by 17 T cell clones from three HLA-DR2/DQw1 homozygous donors are presented in Table VIII. These clones recognized the entire gamut of Hu-BP epitopes, and in contrast to two other studies (35, 36), clonal specificities were not focused on the 87–99 region of BP. Responses in 7 of 7 clones tested were inhibited with anti–HLA-DR antibody, clearly implicating HLA-DR2 as the restriction element, and based on our previous analyses implicating HLA-DR as the predominant locus involved in Hu-BP responses (45) it is likely that the other 10 clones were also DR2 restricted. These data demonstrate that DR2 has the capacity to restrict virtually all of the epitopes of Hu-BP. This highly "permissive" property of the DR2 molecule that allows restriction of many potentially encephalitogenic Hu-BP determinants could contribute to the increased risk of disease associated with this allele.

TABLE VIII

Peptide Specficies of T Cell Clones from MS Patients and Normal Individuals

Donor			Clone number responding to				
Name	Sex	HLA type	P2(1–38)	P1(45–89)	P4(90–170)	None	Total
MS patients							
MS1(NL)	F	DR1,2/DQw1	2	7	0	10	19
MS2(JH)	F	DR7, ?[a]/DQw2,3	3	0	0	1	4
MS3(MD)	F	DR2,4/DQw1,3	2	0	8	0	10
MS4(SO)	F	DR2,w6/DQw1	0	0	1	0	1
MS5(BS)	M	DR2/DQw1	0	1	3	0	4
MS6(MR)	M	DR2/DQw1	3	0	4	3	10
MS7(RB)	M	DR2,7/DQw1	0	1	1	0	2
			10 (20)[b]	9 (18)[b]	17(34)	14(28)[b]	50 (100)
MS8(LB)	F	DR4/DQw3					
MS9(SS)	F	DR5,w6/DQw1,3					
MS10(MB)	F	DR5,w6/DQw1,3					
MS11(JS)	M	DR2/DQw1					
Normal patients							
N1(BP)	F	DR1,3/DQw1,2	0	0	4	1	5
N2(MA)	F	DR3,w6/DQw1,2	2	0	16	2	20
N3(LT)	F	DR2,7/DQw1,2	0	1	2	4	7
N4(DB)	M	DR2/DQw1	0	0	1	2	3
N5(HY)	M	DR4,7/DQw2,3	0	0	1	0	1
N6(JT)	M	DR7/DQw2	0	0	0	23	23
			2 (3)	1 (2)	24(41)	32(54)	59 (100)
N7(SO)	F	DR1,2/DQw1					
N8(PP)	F	DRw8/DQw3					
N9(LS)	F	Not tested					

[a]Antigens in linkage disequilibrium with DR antigens are present, but expected antigens did not type clearly.

[b]Values in parentheses equal % of total clones; compared to normals, $p < .01$.

T Cell Receptor V Region Gene Use

In MS patients, the only way to test the hypothesis that BP- (or any) specific T cells play a pathogenic role in the disease is to regulate the BP-specific T cells and evaluate if there is a clinical effect. It would seem feasible to test this hypothesis for the first time using TCR peptides if MS patients utilize common TCR V region genes. To this end, the T cell clones described above were evaluated for expression of TCR V_β genes by cytofluorometric analysis using a partial set of V gene–specific antibodies (48) and by Northern blot analysis of cDNA amplified from clonal mRNA by the polymerase chain reaction (52) (data kindly provided by Dr. Brian Kozin, Denver, CO).

Hu-BP–specific T cell clones did indeed utilize a limited set of V region genes. As shown in Table IX, $V_\beta 5.2$ was the predominant V region gene, being expressed by 18 of 24 T cell clones from three of the first four MS patients tested. Of the remaining 6 clones, 2 expressed $V_\beta 6.1$. Of note, 13 of 14 clones from MS-1 and 4 of 8 clones from MS-2 expressed $V_\beta 5.2$. These clones were not derived from a single progenitor, since at least 4 different Hu-BP epitopes were recognized by individual $V_\beta 5.2^+$ clones from each patient. In the normal donors, 12 of 12 clones from N-1 expressed $V_\beta 14$.

These data demonstrate clearly that, as in rats and mice, the human T cell repertoire in response to BP is skewed to express a limited set of TCR V region genes. The principles that govern the choice of V genes are not known. However, it is apparent from our data that the overexpression of $V_\beta 5.2$ in MS patients and of $V_\beta 8.2$ in Lewis rats (10, 16) is not epitope driven, since T cells recognizing different epitopes of BP utilized the same TCR V gene. These data also argue against using TCR peptides from CDR3 for inducing antiidiotypic responses, since this hypervariable region has been postulated to interact with the antigenic epitope of the peptide/MHC complex (20, 21) and thus would be expected to differ in primary sequence according to epitope specificities. It is conceivable that the MHC influences the T cell repertoire. Our very limited data set would not contradict this notion, since all four of the MS T cell donors were HLA-DR2$^+$, and a non-DR2 normal donor overexpressed a different V region gene in response to Hu-BP (Table IX). However, other mechanisms such as "superantigen" effects from chronic exposure to microbial infections or autoantigens may also skew the T cell repertoire (53).

With restricted TCR V gene expression in MS patients and with the proven efficacy of TCR peptides as therapy for EAE, the stage is now set to test the hypothesis for the first time that T cells specific for Hu-BP contribute to the MS disease process. Even if BP is not involved, the approach outlined in this chapter may be applicable to regulate T cell responses to other candidate CNS antigens. Moreover, it is conceivable that the general approach will be useful in other autoimmune diseases that involve T cell responses to organ-specific antigens.

TABLE IX
Human MBP Epitope Specificities in DR2/DQw1 Homozygous Donors

Donor	Clone	MBP	Proliferation (cpm/1000 minus background)			
			P2(1–38)	P1(45–89)	P4(90–170)	Epitope/allele
MS5(BS)	#24(14B7)	3 ± 0	0	3 ± 0	0	P1
	#14(8B3)	19 ± 1	0	0	27 ± 0	P4
	#6(5B8)	49 ± 0	0	0	22 ± 2	149-171/DR2
	#23(11D3)	60 ± 4	0	0	31 ± 2	149-171/DR2
MS6(MR)	#9(5D7)	1 ± 0	1 ± 0	0	0	P2
	#43(5D2)	1 ± 0	2 ± 0	0	0	P2
	#52(8C4)	2 ± 0	1 ± 0	0	0	P2
	#41(4F3)	3 ± 0	0	0	1 ± 0	87-99/DR2
	#48(10B7)	2 ± 0	0	0	6 ± 0	130-149/DR2
	#22(3C6)	1 ± 0	0	0	1 ± 0	149-171/DR2
	#45(3C3)	4 ± 0	0	0	2 ± 1	149-171/DR2
	#40(5D10)	13 ± 0	0	0	0	No Peptide
	#51(7B7)	1 ± 0	0	0	0	No Peptide
	#2(3A2)	3 ± 0	0	0	0	No Peptide
Normal	4(DB)					
	#36(9B4)	4 ± 0	0	0	4 ± 0	110-129/DR2
	#9(7E3)	1 ± 0	0	0	0	No Peptide
	#19(8C5)	2 ± 0	0	0	0	No Peptide

SUMMARY

The common use of TCR $V_\beta 8.2$ gene by encephalitogenic, BP-specific T cells from Lewis rats has allowed for the identification of a synthetic TCR sequence that can stimulate antiidiotypic regulation of the pernicious T cells. Preimmunication with the TCR-V_β-8-39-59 peptide induced complete protection against clinical EAE by stimulating the production of specific T cells and antibodies that independently transferred protective activity. Similarly, the TCR peptide suppressed EAE when given after induction of disease, thus reducing the severity of clinical signs. However, histological lesions were evident in clinically protected rats, although T cell responses to encephalitogenic epitopes of BP were muted. The regulatory T cells apparently recognized a processed form of the TCR expressed on the $V_\beta 8^+$ T cell surface in association with MHC class I restriction molecules in order to mediate regulation by noncytolytic mechanisms. It is probable that the TCR-V_β-8-39-59 sequence represents a natural idiotope involved in the immunoregulation of EAE, since T cells specific for this TCR peptide were induced as a consequence of EAE in the absence of synthetic TCR-V_β8-39-59 peptide. Because the regulatory network directed at the TCR-V_β8-39-59 epitope was already present at onset of clinical signs of EAE, it was possible to trigger a recall response to the TCR peptide in sick animals, resulting in rapid resolution of clinical signs. Such TCR peptide therapy may be useful in human autoimmune diseases such as multiple sclerosis, in which BP responses differ qualitatively and quantitatively from normals. Analysis of MS and normal T cell clones confirm that humans also utilized a limited set of TCR V region genes in response to Hu-BP. Thus, it will now be possible to design clinical trials using appropriate TCR peptides to test the hypothesis that BP is a relevant autoantigen in MS.

ACKNOWLEDGMENTS

The authors wish to thank Lynne Gustafson for assistance in preparing the manuscript and Joe Turner for help in preparing the graphics. This work was supported by the Department of Veterans Affairs and National Institutes of Health grants NS23221, NS23444, NS21466, and BNS-8819483.

REFERENCES

1. Fritz, R. M., M. J. Skeen, C. H-J. Chou, M. Garcia, and I. K. Egorov. (1985) *J. Immunol.* **134,** 2328.
2. Beraud, E., T. Reshef, A. A. Vandenbark, H. Offner, R. Fritz, C.-H.J. Chou, D. Bernard, and I. R. Cohen. (1986) *J. Immunol.* **136,** 511.
3. Zamvil, S. S., P. A. Nelson, D. J. Mitchell, R. L. Knobler, R. B. Fritz, and L. Steinman. (1985) *J. Exp. Med.* **162,** 2107.
4. Offner, H., S. W. Brostoff, and A. A. Vandenbark. (1986) *Cell. Immunol.* **100,** 364.
5. Vandenbark, A. A., T. Gill, and H. Offner. (1985) *J. Immunol.* **135,** 223.
6. Bourdette, D. N., A. A. Vandenbark, C. Meshul, R. Whitham, and H. Offner. (1988) *Cell. Immunol.* **112,** 351.
7. Zamvil, S. S., D. M. Mitchell, A. C. Moore, K. Kitamura, L. Steinman, and J. B. Rothbard. (1986) *Nature* **324,** 258.
8. Sakai, K., S. S. Zamvil, D. J. Mitchell, M. Lim, J. B. Rothbard, and L. Steinman. (1988) *J. Neuroimmunol.* **19,** 21.
9. Alvord, E. C., Jr. (1984) In *Experimental Allergic Encephalomyelitis: A Useful Model for Multiple Sclerosis*, vol. 146 (ed. E. C. Alvord, Jr., M. W. Kies, and A. J. Suckling), Alan R. Liss, New York, 523–537.
10. Burns, F. R., X. Li, N. Shen, H. Offner, Y. Chou, A. A. Vandenbark, and E. Heber-Katz. (1989) *J. Exp. Med.* **169,** 27.
11. Acha-Orbea, H., D. J. Mitchell, L. Timmermann, D. C. Wraith, G. S. Tausch, M. K. Waldor, S. S. Zamvill, H. O. McDevit, and L. Steinman. (1988) *Cell* **54,** 263.
12. Chluba, J., C. Steeg, A. Becker, H. Wekerle, and J. T. Epplen. (1989) *Eur. J. Immunol.* **19,** 279.

13. Heber-Katz, E., and H. Acha-Orbea. (1989) *Immunol. Today* **10**, 164.
14. Acha-Orbea, H., L. Steinman, and H. O. McDevitt. (1989) *Annu. Rev. Immunol.* **7**, 371.
15. Kumar, V., D. H. Kono, J. L. Urban, and L. Hood. (1989) *Annu. Rev. Immunol.* **7,**657.
16. Offner, H., G. A. Hashim, B. Celnik, A. Galang, X. Li, F. R. Burns, N. Shen, E. Heber-Katz, and A. A. Vandenbark. (1989) *J. Exp. Med.* **170**, 355.
17. Vandenbark, A. A., Hashim, G. and Offner, H. (1990) In *Dynamic Interactions of Myelin Proteins*, Alan R. Liss, New York, 93–108.
18. Ben-Nun, A., I. R. Cohen. (1981) *Eur. J. Immunol.* **11**, 195.
19. Offner, H., R. Jones, B. Celnik, and A. A. Vandenbark. (1989) *J. Neuroimmunol.* **21**, 13.
20. Claverie, J. M., Prochnica-Chalufour, A., and Bougueleret, L. (1989) *Immunol. Today* **10**, 10–14.
21. Davis, M. M. and Bjorkman, P. M. (1988) *Nature* **334**, 395–402.
22. Margalit, H., Spouge, J. L., Cornette, J. L., Ease, K. B., Delisi, C., and Berzofsky, J. A. (1987) *J. Immunol.* **138**, 2213–2219.
23. Rothbard, J. B., and Taylor, W. R. (1988) *EMBO J.* **7**, 93–100.
24. Vandenbark, A. A., G. Hashim, and H. Offner. (1989) *Nature* **341**, 541.
25. Williams, C. B. and Gutman, G. A. (1989) *J. Immunol.* **142**, 1027–1035.
26. Howell, M. D., S. T. Winters, T. Olee, H. C. Powell, D. J. Carlo, and S. Brostoff. (1989) *Science* **246**, 668.
27. Lyman, W. D., Abrams, G. A., and Raine, C. S. (1989) *J. Neuroimmunol.* **25**, 195.
28. Bourdette, D. N., Vainiene, M., Morrison, W. J., Jones, R., Turner, M. J., Vandenbark, A. A., and Offner, H. (1991) *J. Neurosci. Res.* **30**, 308.
29. Schluesener, H. J., and Lider, O. (1989) *J. Neuroimmunol.* **24**, 249.
30. Karpus, W. J., and Swanborg, R. J. (1990) *FASEB J.* **4**, A1856–948.
31. Hashim, G. A., Vandenbark, A. A., Galang, A. B., Diamanduros, T., Carvalho, E., Srinivasan, J., Jones, R., Vainiene, M., Morrison, W. M., and Offner, H. (1990) *J. Immunol.* **144**, 4621–4627.
32. Urban, J., V. Kumar, D. Kono, C. Gomez, S. J. Horvath, J. Clayton, D. G. Ando, E. E. Sercarz, and L. Hood. (1988) *Cell* **54**, 577.
33. Owhashi, M. and E. Heber-Katz. (1988) *J. Exp. Med.* **168**, 2153.
34. Oksenberg, J. R., S. Stuart, A. B. Begovich, R. B. Bell, H. A. Erlich, L. Steinman, and C. C. A. Bernard. (1990) *Nature* **345**, 344.
35. Wucherpfennig, K. W., K. Ota, N. Endo, J. G. Seidman, A. Rosenzweig, H. S. Weiner, and D. A. Haffler. (1990) *Science* **248**, 1016.
36. Ota, K., M. Matsui, E. L. Milford, G. A. Mackin, H. L. Weiner, and D. A. Hafler. (1990) *Nature* **346**, 183–187.
37. Paterson, P. Y., (1966) *Adv. Immunol.* **5**, 131.
38. Waksman, B. H., and Reynolds, W. E. (1984) *Proc. Soc. Exp. Biol. Med.* **175**, 282.
39. Burns, J., Zweiman, B., and Lisak, R. P. (1983) *Cell. Immunol.* **81**, 435.
40. Tournier-Lasserve, E., Hashim, G. A., and Bach, M. A. (1988) *J. Neurosci. Res.* **19**, 149.
41. Vandenbark, A. A., Chou, Y. K., Bourdette, D., Whitham, R., Chilgren, J., Chou, C. H.-J., Konat, G., Hashim, G., Vainiene, M., and Offner H. (1989) *J. Neurosci. Res.* **23**, 21.
42. Hafler, D. A., Benjamin, D. S., Burks, J., and Weiner, H. L. (1987) *J. Immunol.* **139**, 68.
43. Richert, J. R., Robinson, E. D., Deibler, G. E., Martenson, R. E., Dragovic, L. J., and Kies, M. W. (1989). *J. Neuroimmunol.* **23**, 55.
44. Zhang, J., Chou, C. H.-J., Hashim, G., and Raus, J. (1990) *Cell Immunol.* **129**, 189.
45. Chou, Y. K., Vainiene, M., Whitham, R., Bourdtee, D., Chou, C. H.-J., Hashim, G., Offner, H., and Vandenbark, (1989) *J. Neurosci. Res.* **23**, 207.
46. Pette, M., Fujita, K., Giegerich, G., Trowsdale, J., Kappos, L., and Wekerle, H. (1990) *Neurology* **40** (Suppl 1), 391.

47. Hauser, S. L., Fleischnick, E., Weiner, H. L., Marcus, D., Awdeh, Z., Yunis, E. J., and Alper, C. A. (1989) *Neurology* **39,** 275.

48. Chou, Y. K., Henderikx, P., Vainiene, M., Whitham, R., Bourdette, D., Chou, C.-H. J., Hashim, G., Offner, H., and Vandenbark, A. A. (1991) *J. Neurosci. Res.* **28,** 280.

49. Allegretta, M., Nicklas, J. A., Sriram, S., and Albertini, R. J. (1990) *Science* **247,** 718.

50. Link, H., Olsson, T., Wang, W. Z., Jiang, Y. P., and Höjeberg, B. (1990) *Neurology* **40** (Suppl 1), 283.

51. Panitch, H. S., Haley, A. S., Hirsch, R. L., and Johnson, K. P. (1986) *Neurology* **36,** 285.

52. Francis, D. A., Batchelor, J. R., McDonald, W. I., Hern, J. E. C. and Downie, A. W. (1986) *Lancet* **1,** 211.

53. Choi, Y., Kozin, B., Herron, L., Callahan, J., Marrack, P., and Kappler, J. (1989) *Proc. Natl. Acad. Sci. USA* **86,** 8941.

54. Kappler, J., Kotzin, B., Herron, L., Gelfand, E. W., Bigler, R. D., Boylston, A., Carrel, S., Posnett, D. N., Choi, Y., and Marrack, P. (1989) *Science* **244,** 811–813.

PART VI

PERSPECTIVES

22

Tolerance of Self: Present and Future[1]

N. A. MITCHISON
Deutsches Rheuma Forschungszentrum Berlin
D-1000 Berlin 39, Germany

New information is presented in this volume showing (i) how B cells develop, including the stages that they pass through in generating molecular diversity; (ii) from studies on V gene usage, and especially from transgenes, that positive as well as negative selection really do occur in the thymus, and something about when and where they do so; (iii) how T and B cells handle positive signals, and something also about how they may handle negative signals; (iv) how a first encounter with antigen can engender a state of anergy in a postthymic T cell; (v) how logically to design peptide-based therapy for autoimmune diseases; and (vi) how to design therapy using monoclonal antibodies. These are impressive advances, and for that very reason it is easy to overlook progress made in the previous era. Accordingly, this overview begins with a retrospective glance at the baseline from which this recent progress started. The state of play at that time is illustrated in a review written at that time (1).

THE 1975 BASELINE

At that time immunological tolerance was understood in the following terms.

1. Self-tolerance results from the deletion of self-reactive T cells in the thymus. For reasons which are discussed in greater detail below in connection with the F antigen story, it was already known that self-tolerance among B cells was a secondary issue. The conclusion that it results from deletion was drawn from negative evidence. Various suppressor mechanisms had been discovered

[1]Based on the concluding address given at the P&S Biomedical Sciences Symposium, "Molecular Mechanisms of Immunological Self-Recognition."

in manipulated systems, for instance, in rats that had been made tolerant by neonatal injection of allogeneic cells, but none of these could be made manifest in unmanipulated systems, notably again with the F antigen. The more was learned about suppression in later years, the clearer it became that this kind of activity was associated not with self-tolerance itself, but rather with the autoimmunity that followed breakdown of self-tolerance. And the more was learned about it, the more meaningful became the negative findings in unmanipulated self-tolerance. This is not to detract from the importance of the direct demonstration of clonal deletion that came from V gene usage studies; that is beautiful science, but the results were not unexpected. Indeed, the very fact that the V region antibodies could be used for this purpose goes a long way to justify the great efforts devoted to their production.

2. Although the thymus is evidently the main natural site for the acquisition of tolerance, it was known that it could also be acquired by postthymic T cells. Indeed, from the very first study made with a foreign serum protein in the mouse (bovine immunoglobulin), it was clear that the mature immune system was fully able to respond to a foreign antigen by becoming tolerant. Furthermore, it was known that when a foreign antigen was administered under conditions which would lead to the induction of tolerance, it made no detectable difference to the outcome whether the thymus was present or not.

3. The conditions that would lead to tolerance had become fairly clear. The key was somehow to administer antigen without provoking a positive immune response. This could be achieved in various ways, such as by starting treatment neonatally, or by injecting adults with aggregate-free foreign serum proteins. Some routes of administration proved more effective than others, for reasons that have never been entirely clear. For instance, oral administration is particularly effective, possibly because the liver can filter off particles before they reached lymphoid tissue. Another useful maneuver was to damage the immune system by antilymphocyte serum or by irradiation. Not much distinction was made at that time between depleting and "nondepleting" treatments; the latter came in with cyclosporin A, a form of treatment that in effect blindfolds but does not destroy T cells. So although Waldmann's tolerization under cover of monoclonal anti-CD4 opens up new prospects in therapy, as described in his preceding chapter, the science is not unexpected.

From all these findings, the negative (tolerance) response came to be accepted as the default condition, that would ensue whenever a positive response did not occur.

4. A wide variety of soluble proteins, including nonserum proteins such as ovalbumin, proved able to induce tolerance at a highly consistent threshold concentration in body fluid of between 10^{-8} and $10^{-9} M$. It made no detectable difference to this threshold whether induction took place in the thymus or in mature lymphoid tissue. This seemed to mean something, and one idea at the

time was that this might reflect the lowest concentration at which T cells could detect antigen in solution. That is no longer tenable, of course, as it contradicts the Bjorkman–Wylie paradigm. In any case this idea did not fit with another observation of that time, that T cells would become tolerant only on passage through lymphoid tissue, and not if simply left in the blood. What all this means in terms of antigen presentation is still not known, although as the cell biology of antigen processing progresses it may well become clear.

5. The continued presence of antigen was known to be required for the maintenance of tolerance, unless the thymus had been removed. This is entirely logical, as freshly matured T cells would otherwise bring back immunological reactivity. Furthermore, the speed with which tolerance wanes was found to be proportional to age, presumably reflecting the gradual waning of thymus activity.

These observations played an important part in the immunological theories of the day. They led Medawar, for instance, to toy with the possibility that the tolerant state might somehow resemble substrate adaptation in bacteria, an idea that he soon abandoned when clonal selection swept the board. Later, when T cells with suppressor activity were discovered and the question was asked whether these might play a role in self-tolerance, these observations provided cogent arguments against that possibility.

So what was missing in this earlier picture of self-tolerance? Conspicuous for its absence was any real understanding of the cell biology. There was little idea about either the signals or the intracellular mechanisms that led to the elimination of a self-reactive lymphocyte. Clearly, immaturity per se was not a crucial factor for T cells, although it might be for B cells. Systems in which experimentation could be carried out were lacking, particularly *in vitro* ones, and that is where recent progress has been most encouraging.

F LIVER ANTIGEN

Work on F liver antigen in my laboratory did much to support these views (2). This 42-kDa protein is highly conserved in vertebrates. It is of immunological interest because (i) it occurs in body fluid at a concentration ($10^{-9} M$) low enough to tolerize T but not B cells, (ii) it has in mice an allo-variable site that can be recognized by T cells (the gene encoding this protein in mice has now been cloned in the laboratory of D. Oliveira (Cambridge University), and the alloepitope provisionally identified), and (iii) in liver it occurs at concentrations high enough to make it easily extractable. The consequence of all this is that mice can readily be immunized against allo-F protein, and they then make antibodies which recognize allo- and self-F equally well. This itself provides perhaps the best available evidence for a difference between T and B cells in susceptibility to tolerization. Furthermore, T cells able to suppress the anti-F response have been demonstrated (as also have the class II major histocompatibility molecules that they recognize), but they play no part in tolerance of self as it occurs naturally.

NEGATIVE SIGNALING

The two contributions to this volume which deal most directly with the issue of negative signaling, are those of Perlmutter and Schwartz. Neither of these have final results for us yet, but excitingly they illustrate how the full force of molecular biology can be brought to bear on the problem. Much the same could be said of the current work of J. J. T. Owen and R. Mishell in Birmingham and K. Okomura and colleagues in Tokyo. Both of these groups have focused effort on the negative response of thymocytes to treatment with anti-CD3 monoclonal antibodies. Tessa Crompton has demonstrated the same effect *in vivo* in my laboratory, by administering anti-CD3 antibodies to mice that have been heavily prelabeled with I^{125}-labeled uridine deoxyriboside.

Good luck to the molecular biologists! However, there are catches that they need to be aware of. One applies to incomplete signals (i.e., antigen without costimulators) inducing anergy, as discussed above by Schwartz. To make this a convincing *in vitro* model of self-tolerance, it will eventually need to fit into the quantitation mentioned above. That means that it will need to be inducible by proteins or peptides in the concentration range of 10^{-8} to $10^{-9} M$, and that is far below the concentrations currently in use.

Another catch applies to anti-CD3–induced cell death in the thymus. For this to provide a convincing model of physiological events, it should operate at the stage of development where T cells normally undergo negative selection. This volume includes somewhat differing conclusions on just when that is. Guidos and Weissman (this volume) argue that it occurs late, at the end of the CD4CD8 double positive stage of T cell development, while the transgenic studies indicate that it occurs earlier. This discrepancy may simply reflect the relatively early stage at which receptors encoded by transgenes get expressed. The catch is connected with expression of the CD45 isoforms. The majority of thymocytes in the mouse are CD45R⁻, and these are the cells mainly affected by anti-CD3 treatment. However, the work of Lightstone and Marvel cited below suggests that this may not be the population within which negative selection normally occurs. The implication of this is that anti-CD–induced death may represent only the acceleration of a developmental program leading to apoptosis, to which these cells are already committed, rather than a valid model of negative selection.

ANERGY, AND THE NEED FOR A DEFINED
ANERGIC PHENOTYPE

Schwartz describes the hunt which he and Lenardo are conducting for the molecular lesion in the anergic T cell, and Goodnow describes a preliminary characterization of the anergic B cell in terms of its localization in lymphoid tissue. One can be optimistic about the outcome of the molecular hunt, simply because somewhere further up the transcriptional control cascade there must be a lesion to be found. Studies on transcriptional control factors are moving so fast, and so unpredictably, that the lesion may be disclosed at any time. Another reason for optimism is that these hunters are not alone. The group of my colleague Martin Raff, for instance, finds that a time comes in development when the O2A progenitor (the cell which gives rise to oligodendrocytes and type II astrocytes) can no longer respond to platelet-derived growth factor. Yet, like Schwartz's anergic T cell, it retains its receptors and second messenger response intact: the parallel is close.

The exciting time will come when the anergic phenotype has been sufficiently well characterized to make an *in vivo* assessment possible. Only then could it be discovered whether anergy plays a significant part in the natural acquisition of self-tolerance.

THYMUS LOBE CULTURES

In the search for an *in vitro* system of tolerance induction, the Owen–Jenkinson thymus lobe culture has much to offer. A tiny fetal lobe that contains less than 10^5 precursors and no mature T cells grows over a culture period of 10 days to attain a content of 10^6 T cells, of which some 10% have the mature phenotype. This system my colleagues and I have been exploring in several directions. We have asked whether CD45RA is a lineage marker (3). As mentioned above, the majority of thymocytes lose this isoform and differentiate into CD45R expression. Within the earliest population CD4CD8 double negative cells, only the CD45RA cells have progenitor activity. Within the next population, of CD4CD8 double positives, E. Lightstone and J. Marvel have identified a small population of CD45RA blasts. They do not yet know whether progenitor activity is confined to this population, and although this needs to be done it is a daunting task. But at any rate it is encouraging to find that this population can be detected, after its existence had been postulated ("hypothesis 2" in ref. 3).

Another promising use for the mouse lobe cultures has been as a site for growing human thymocytes and enabling them to differentiate. This A. Fisher, M. Merkenschlager, and their colleagues have managed to do, using not only fresh fetal thymocytes but also ones that have been preserved in the frozen state. This provides a powerful system in which to explore the impact of various anti-CD monoclonal antibodies on the survival and development of thymocytes.

Katharina Simon and I have found that the mouse T cells which mature in these cultures do not become tolerant of self-F protein, for upon return to syngeneic hosts they can respond to immunization with this protein in adjuvant. We shall attempt to tolerize these cells within the cultures, so as to find out more about tolerization in an *in vitro* system. An immediate application of such a system would be to examine the paradox that T cells are normally exposed to sufficiently high concentrations of F protein to become tolerant, even though freshly isolated presenting cells do not retain enough of the protein to drive the proliferation of primed T cells (3). Figure 1 provides a guide to current thinking about access of self-macromolecules to the thymus. It refers to the four categories of molecules that can induce self-tolerance: those which are made in the thymus, those which are made outside but reach the thymus in amounts sufficient to be readily detectable on presenting cells, those which reach the thymus in lesser amounts, and those which perhaps do not reach the thymus at all. We have little idea of the relative importance of these four categories, and there is much to discover about just how they affect T cell development.

AUTOIMMUNITY AND EPITOPE LINKAGE

A series of studies on hapten-conjugated proteins and on cellular antigens have demonstrated the importance of epitope linkage in coordinating the immune response (3, 4). This is a mechanism implemented by regulatory T cells, which causes the response to spread from one epitope to another throughout an antigenic structure. The structure can be larger than a single protein, such as a virion or a group of proteins inserted into the same cell membrane, provided that it is processed in the immune system as a single unit.

The groups of marker antibodies that occur together in lupus erythematosis, dermatomyositis, and other autoimmune diseases provide striking examples of this phenomenon. Figure 2 summarizes information provided about these antibodies in a recent review (5) and also includes (at the top) information from my own laboratory concerning membrane antigens (3). This is not the end of the story: in most cases the regulatory T cells that mediate these linkages have not yet been identified, and just how these groups of proteins and nucleic acids are coprocessed is not known.

Very similar views about rheumatoid factors have recently been expressed independently by two groups (E. Roosnik and A. Lanzavecchia, and J. Gergeyi and colleagues, both as personal communication). The hypothesis is that any B

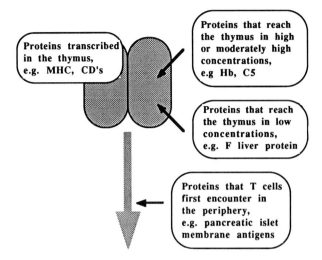

Fig. 1.   Access of self-macromolecules to the thymus.

cell reactive with self-immunoglobulin could take up microbial antigens in the form of immune complexes and would then receive help from regulatory T cells reactive with those antigens. Such a mechanism could well explain the rheumatoid factors which are often made in the course of infection and possibly also those of rheumatoid arthritis itself.

TOWARD GENE THERAPY IN AUTOIMMUNITY

Yet another interesting finding made with F antigen is the first single allele substitution that profoundly inhibits an antibody response (6). Whether the inhibition operates through negative selection (deletion of expressed V genes) or positive selection (development of inhibitory T cells) is not known. McDevitt and Nepom (7) have dealt with the same issue in the context of the controversial protective effect mediated by the Asp-57 substition in HLA-DQ (7). In fact inhibitory class II HLA genes must be quite common, to judge by the frequency with which negative associations turn up during HLA surveys in autoimmune diseases such as insulin-dependent diabetes and rheumatoid arthritis. Having a strongly inhibitory gene in the mouse should facilitate studies of mechanism.

In this volume, Hämmerling mentions a transgenic mouse that expresses in a highly restricted manner H-2A^b, the very gene that Simon and I have identified

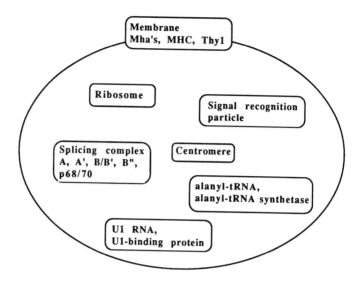

Fig. 2. Groups of cellular antigens that appear to be coprocessed by the immune system.

as able to inhibit the anti-F response. We hope that this will lead to a collaborative study, in which we shall titrate the minimum expression of a class II major histocompatibility complex gene needed to produce inhibition.

The message from the major histocompatibility transgene studies in this volume is (i) that these molecules can profoundly perturb the immune system, even when they are expressed at low levels, and (ii) that they can do so whether or not they are expressed in the thymus itself. Where will this take us? Before long we shall see the first trials of gene therapy in humans probably in cancer and in congenital enzyme deficiencies. No doubt they will be followed by trials in the hemoglobinopathies. If these go well, it would seem reasonable to consider such therapy for autoimmunities which have proved refractory to other forms of treatment. Of course, there are difficulties: how to protect an implanted major alloantigen, for example. But with the technologies that are becoming available none of them seem insuperable. Now may be the time to start a serious research effort toward that goal.

REFERENCES

1. Howard, J. G., and Mitchison, N. A. (1975) *Prog. Allergy* **18**, 43–96.
2. Griffiths, J. A., Mitchison, N. A., Nardi, N., and Oliveira, D. B. G. (1988) In *Immunogenicity of Protein Antigens* (ed. E. Sercarz and J. Berzofsky), CRC Press, Boca Raton, FL., 35–41.
3. Fisher, A. G., Goff, L. K., Lightstone, E., Marvel, J., Mitchison, N. A., Poirier, G., Stauss, H., and Zamoyska, R. (1989) *Cold Spring Harbor Symp. Quant. Biol.* **54,** 667–674.
4. Lake, P., Mitchison, N. A., Clark, E. A., Khorshidi, M., Nakashima, I., Bromberg, J. S., Brunswick, M. R., Szensky, T., Sainis, K. B., Sunshine, G. H., Favilla-Castillo, L., Woody, J. N., and Lebwohl, D. (1989) In *Cell Surface Antigen Thy-1* (ed. A. E. Reif and M. Schlesinger), Marcel Dekker, New York, 367–394.
5. Plotz, P. H. (1990) In *The Role of Micro-Organisms in Non-Infectious Diseases.* (ed. R. R. P. de Vries), Springer, London, 111–121.
6. Mitchison, N. A., and Simon K. (1990) *Immunogenetics* **32,** 104–109.
7. Nepom, G. T. (1990) *Immunol. Today* **11,** 314–315.

Index

ISBN 0-12-053750-8

90050

9 780120 537501